Artificial Intelligence-Enhanced Software and Systems Engineering

Volume 5

Series Editors

Maria Virvou, Department of Informatics, University of Piraeus, Piraeus, Greece

George A. Tsihrintzis, Department of Informatics, University of Piraeus, Piraeus, Greece

Nikolaos G. Bourbakis, College of Engineering and Computer Science, Wright State University, Joshi Research, Dayton, USA

Lakhmi C. Jain, KES International, Shoreham-by-Sea, UK

T0181791

The book series AI-SSE publishes new developments and advances on all aspects of Artificial Intelligence-enhanced Software and Systems Engineering—quickly and with a high quality. The series provides a concise coverage of the particular topics from both the vantage point of a newcomer and that of a highly specialized researcher in these scientific disciplines, which results in a significant cross-fertilization and research dissemination. To maximize dissemination of research results and knowledge in these disciplines, the series will publish edited books, monographs, handbooks, textbooks and conference proceedings. Of particular value to both the contributors and the readership are the short publication timeframe and the world-wide distribution, which enable both wide and rapid dissemination of research output.

Christophe Gaie · Mayuri Mehta
Editors

Recent Advances in Data and Algorithms for e-Government

 Springer

Editors
Christophe Gaie
Centre Interministériel de Services
Informatiques Relatifs aux Ressources
Humaines (CISIRH)
Paris, France

Mayuri Mehta
Department of Computer Engineering
Sarvajanik College of Engineering
and Technology
Sarvajanik University
Surat, Gujarat, India

ISSN 2731-6025 ISSN 2731-6033 (electronic)
Artificial Intelligence-Enhanced Software and Systems Engineering
ISBN 978-3-031-22410-2 ISBN 978-3-031-22408-9 (eBook)
https://doi.org/10.1007/978-3-031-22408-9

This Springer imprint is published by the registered company Springer Nature Switzerland AG
The registered company address is: Gewerbestrasse 11, 6330 Cham, Switzerland

Preface

Nowadays, data are the core of information systems and constitute a significant asset as they offer insight on the system's status and maturity. Data make a decisive contribution to valourizing private activity, as observed daily through the most powerful companies such as Google, Apple, Facebook, Amazon, and Microsoft (i.e. the GAFAM). They facilitate trade development, make flourishing advertising and entertainment, or enhance technology in favour of customers. Their power and strength are growing with a high velocity thanks to data valourization (especially information concerning customers).

However the interest towards data offers companies an immediate return on investment, obtaining benefits is more challenging for government organizations. Indeed, government services do not provide profitable services but rather fulfil the needs of the general interest, such as education, healthcare, safety, or security. Moreover, government administrations offer better protection of personal data rights, reducing their ability to exchange and combine data from different origins. The orientation towards collecting the explicit consent of citizens is a priceless guarantee of democracy as well as a complexity to provide customized services for better efficiency.

These first considerations legitimate the following interrogations: How to take advantage of data to provide better public services? May algorithms be helpful to valourize data and contribute to the general interest? Should government services be inspired by private services and/or expert government organizations?

An interesting concern to facilitate the development of better public web services relies on data reusability. It aims to define how to convert data extracted from an existing information system into data that is easy to manipulate and reuse in any context. Organizations tackle this subject through the prism of the data life cycle. Indeed, IT organizations evolve progressively from producing raw data, then organizing data to ensure a better organization and finally optimizing its usage between internal services and/or by final customers.

In the context of e-Government, the focus towards data reusability aims to fulfil different objectives. First, governments aim to enhance data quality to offer the highest preciseness of public services, which contributes to ensuring 'citizens' satisfaction, equality of treatment, and regularity of procedures. Then, the ability to reuse data aims to reduce administrative burdens by introducing a single entry point instead of multiple shopping procedures. This may rely on the once-only-principle (OOP) or other back-office data exchange systems. Finally, data for multiple administrative proceedings reduces public service costs for better efficiency.

The current book tackles recent research concerning the rise of e-Government which largely combines two dimensions: the valourization of data that government administrations hold and the conception of algorithms to increase public services. The book contains multiple use cases to illustrate the modernization of government services. The editors are thankful to the authors who submitted their research work to this book as well as to all the anonymous reviewers for their insightful remarks and significant suggestions. We hope readers will find the book useful and be inspired to contribute to government modernization.

Paris, France   Christophe Gaie
Surat, India   Mayuri Mehta
October 2022

Contents

Abbreviations

3D	Three dimensional
ACT	American College Test
Agri	Agriculture
AI	Artificial intelligence
AMS	Assessment management system
API	Application programming interface
ARS	Agriculture Research Service
AUP	Acceptable use policy
B2C	Business to consumer
BESCOM	Bangalore Electricity Supply Company Limited
BMRCL	Bangalore Metro Rail Corporation Limited
BMTC	Bengaluru Metropolitan Transport Corporation
BWSSB	Bangalore Water Supply and Sewerage Board
BYOD	Bring your own device
C2C	Citizen-to-citizen
CCOPM	Capacity-capability-opportunity-potential Model
CCSS	Common core standards
CCT	Communication and collaboration tools
CE	Collaborative Enterprise
CGU	Brazilian Office of the Comptroller General
CIOs	Chief information officers
CITC	Communications and Information Technology Commission
CL	Culture of literacy
ConecteSUS	ConecteSUS system
CoSN	Consortium for School Network
COVID-19	Coronavirus Disease 2019
CRM	Customer relationship management
CSC	Common Service Centre
DAERA	Department of Agriculture, Environment, and Rural Affairs
DCL	District, Community or Locale
DETA	Distance Education and Technological Advancements

DHC	Descending Hierarchical Classification
DHS	Department of Homeland Security
DIA	Direct Internet Access
DIAS	Data and Information Access Services
DSFAS	Agriculture and Food Research Initiative's Data Science for Food and Agricultural Systems
EA	Enterprise architecture
EC	European Commission
EDSS	Electronic Document Delivery System
EGDI	The United Nations E-Government Development Index
e-Gov	Electronic Government
E-Governance	Electronic Governance
E-Government	Electronic Government
EIN	Education Information Network
EISSN	Electronic International Standard Serial Number
EMDS	Ecosystem Management Decision Support System
EMIS	Education management information systems
EPI	E-participation index
EPIS	Korea Agency of Education, Promotion and Information Service in Food, Agriculture, Forestry and Fisheries
ERS	Economic Research Service
e-service	Electronic service
EU	European Union
FAA	Federal Aviation Administration
FAC	Factor analysis of correspondence
FAO	Food and Agriculture Organization of the United Nations
FCA	Factorial correspondence analysis
FEA	Federal enterprise architecture
FFP	Filtering face piece particles
FSA	Farm Service Agency
G2B	Government to business
G2C	Government to consumer or government to citizen (according to the context)
G2E	Government to employee
G2G	Government-to-government
GBAD	Graph-based anomaly detection
GDP	Gross domestic product
GDS	Government Digital Service
GOI	Government of India
Green IT	Green Information Technology
GST	Goods and Services Tax
ICRISAT	International Crops Research Institute for the Semi-Arid Tropics
ICT	Information and Communication Technology
ICT-CFT	ICT Competency Framework for Teachers
ID	Identity document

IIA	Israel Innovation Authority
IITE	Institute of Information Technologies in Education
ILN	Indonesia literacy network
IMF	International Monetary Fund
IndEA	India Enterprise Architecture Framework
IoB	Internet of Behaviour
IoF	Internet of Farming
IoT	Internet of Things
IP	Internet protocol
IS	Information systems
ISSN	International Standard Serial Number
ISTE	International Society for Technology in Education
IT	Information Technology
ITSM	Information Technology Service Management
ITU	International Telecommunication Union
KPI	Knowledge performance indicator
LCAT	Land change analysis tool
LCMS	Landscape change monitoring system
LGPD	Brazilian General Data Protection Law (in Portuguese)
LMS	Learning management system
LP	Literacy posts
MAFRA	Ministry of Agriculture, Food and Rural Affairs
ML	Machine learning
MLA	Member of Legislative Assembly
MLP	Multilayer perceptron
MoE	Minstry of Education
MoEP	Ministry of Economy and Planning
MOOCs	Massive open online courses
MP	Member of Parliament
mRNA	Messenger ribonucleic acid
MS	Microsoft
NAII	National Artificial Intelligence Initiative
NAL	National Agricultural Library
NARO	National Agriculture and Food Research Organization
NCEPD	National Center for Educational Professional Development
NFC	Near-field communication
NGO	Non-Government Organization
NIFA	National Institute of Food and Agriculture
NLCD	National Land Cover Database
NLI	National Library of Indonesia
NLP	Natural language processing
NPM	New public management techniques
NSF	National Science Foundation
NTP	National Transformation Plan
OECD	Organization for Economic Co-operation and Development

OLC	Online Learning Consortium
PA	Perpusnas application
PAN	Permanent account number
PDS	Public Distribution System
PERPUSNAS	Perpustakaan Nasional
POER	Perceived organizational e-readiness theory
PVT	Process virtualization theory
QALY	Quality-adjusted life year
QM	Quality matters
R&D	Research and development
RI	Republic Indonesia
RPT	Related party transactions
RTO	Road Transport Office
SABER	Systems Approach for Better Education Results
SAT	Scholastic Assessment Test
SDG	Sustainable Development Goals
SEIR	Susceptible exposed infectious recovered
SEND	Smart Education Networks by Design
SFA	Singapore Food Agency
SHEILA	Supporting Higher Education to Integrate Learning Analytics
SICONFI	Public sector accounting and fiscal information system
SIS	Student information system
SLR	Systematic literature review
SOA	Service-oriented architecture
SSO	Single sign-on
SUS	Brazilian National Health System
SUV	Sport utility vehicle
SVIR	Susceptible vaccinated infected recovered
SVM	Support vector machine
T4edu	Tatweer Company for Education Services
TAM	Technology acceptance model
TETCO	Tatweer Educational Technologies Company
TOGAF	The Open Group Architecture Framework
TSM	Twitter social media
UAS	Unmanned aircraft systems
UAV	Unmanned aerial vehicle
UBS	Basic health units
UCE	Elementary context units
UK	United Kingdom
UN	United Nations
UNCRC	United Nations Convention on the Rights of the Child
UNESCO	United Nations Educational Scientific and Cultural Organization
UNICEF	United Nations International Children's Emergency Fund
USA	United States of America
USD	United States Dollar

USDA	United States Department of Agriculture
UTAUT	Unified theory of acceptance and use of technology
VAT	Value-added tax
VC	Venture capitals
VPS	Vaccination and protection strategy
WaPOR	Water productivity through open access of remotely sensed derived data
WCAG	Web Content Accessibility Guidelines
WHO	World Health Organization
XBRL	eXtensible Business Reporting Language

Chapter 1
Conceptual Model and Data Algorithm for Modernization of e-Governance Towards Sustainable e-Government Services

D. Vimala, S. Vasantha, and A. Shanmathi

Abstract The study aims to examine the effect of modernization of electronic government services on government service efficiency and citizen satisfaction to attain sustainable e-government services. The authors proposed conceptual research frame work to examine the determinants of Service Quality of e-governance and how the integration of State-of-the-art technology impact sustainable e government services. Case study approach is adopted to explore best modernization e-governance service delivery models practiced across the globe and development of data algorithm to validate citizen's data. The study is based on comprehensive in-depth literature review. The methodology followed in the study is a multi-faceted approach consists of development of conceptual research framework, case study method and data algorithm. This study scrutinized the relationship between the various attributes that help to attain sustainability of e-government services. The performance of the state's activities can be improved through excellent service quality which is determined by information quality and system quality. The modernisation of e-governance services in various countries proved that there is a positive impact on citizen satisfaction and sustainable e-governance.

Keywords Citizen · Data algorithm · e-Government services · Information quality · Modernization satisfaction · Sustainable e-government services · System quality

1.1 Introduction

The acquisition and utilization of electronic government services is a major challenge in developing countries. Digital gaps in society, a lack of e-government services

D. Vimala · S. Vasantha (✉)
School of Management Studies, Vels Institute of Science, Technology & Advanced Studies (VISTAS), Chennai, India
e-mail: vasantha.sms@velsuniv.ac.in

A. Shanmathi
HCL Technologies, Chennai, India

© The Author(s), under exclusive license to Springer Nature Switzerland AG 2023
C. Gaie and M. Mehta (eds.), *Recent Advances in Data and Algorithms for e-Government*, Artificial Intelligence-Enhanced Software and Systems Engineering 5, https://doi.org/10.1007/978-3-031-22408-9_1

and poor accessibility to technology are serious problems that require immediate attention. As a result, governments are innovating new techniques to make use of the resources efficiently to provide services to citizens. Citizens' hesitation to use an e-government system again after receiving subpar service determines the legality of investments in e-government arrangements and initiatives. The improved level of online services can stimulate e-government adoption by meeting citizen expectations [1].

Researchers and practitioners in this sector have been enthralled by the promise of e-government to increase pellucidity and reduce exploitation in the public administration. It has been stated that e-government in this context symbolizes a "new phase of government," one that is reshaping communication across all tiers of the government [2].

Asia is an example of a region that has been impacted by the electronic services provided by the government. Iqbal and Seo [3], for example, demonstrated that improving e-governance procedures was an effective approach. Electronic government services had a role in the restructuring of Japan's public sector, notably in terms of public accountability. Some experts say Singapore's superior e-government system has accomplished its objective. China's e-governance progress is also significant. The quality of public services suffers significantly when an economy's governance is weak. The ultimate aim of the World Development Report was also service delivery. The fundamental rationale for focusing on service delivery was that, in modern times, the public sector has been slow and insensitive to residents' needs. According to the World Bank and International Bank for Reconstruction and Development, public service delivery in emerging countries has not been consistent with population preferences (2005).

A true citizen-centric public sector can be enabled by a smart and proactive government. Globally it is proved that in many countries the government is trying to enhance citizen access to real-time answers, particularly for routine tasks through the application of artificial intelligence. But this could be improved when modernizing direct two-way interaction between citizens and the state through data and algorithm with the support of state-of-the-art technology to achieve citizen satisfaction.

1.1.1   Evolution of e-Government

With the development of the internet in the 1990s, there were widespread movements toward government IT adoption. Subsequently both technology and e-governance projects have advanced significantly. The number of internet and mobile connections is growing, and people are accessing in a variety of ways. They have begun to anticipate more and more services from corporations and governmental entities. The idea of e-government was first introduced in India in the year 1970s, for the management of data-intensive tasks such as managing elections, censuses, tax administration, and economic monitoring (Fig. 1.1).

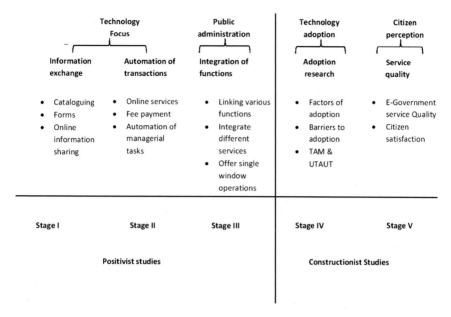

Fig. 1.1 Evolution of e-government

Positivist approach in this study adheres to the view that factual knowledge gained through observation is trust worthy. Positivist study highlights the initial stage of technology focus and public administration which comprises of information exchange, automation of transactions and integration of functions. Since 2020, i.e., post pandemic, the online activities of the citizen emerged to be quiet high. COVID-19 pandemic has increased the usage of e-government services and accelerated the digitalisation as well as trust towards e-government services.

The factors such as Information exchange and integration of all functions are experienced with the human intellects. Scientific verification of information is required in the positivist approach.

The next two stages of evolution are categorized under constructionist studies that consist of Technology adoption and citizen perception. Since the factors influencing adoption and barriers to adoption is identified through human construct. Subsequently e-government citizen quality and citizen satisfaction is measured by Government interaction with citizen for example G2C and vice versa, they come under constructionist studies.

At present, after the pandemic, the adoption of technology is irreplaceable. Digital technologies have made tremendous contributions in addressing social and economic problems faced by the citizens during pandemic as well as expedited two way interactions (Government- citizen) effectively.

Currently, the government states are in the transformational and developmental stage, i.e., the government reform strategy which radically change the way, people understand government and to promote citizen-centric service delivery mechanism.

It is a trend for recent years providing services through one-stop shop all over the world. It is a single point of contact where Governments can collect data through a single, integrated digital form for one or more services.

1.1.2 State-of-the-Art Technology in e-Government Services

Sayabek Ziyadin [4] conducted a study to prove that big data has made it possible to integrate e-government into government operations on a systemic level through data collaboration. Improvements in public services, enhanced institutional effectiveness, political involvement and transparency, as well as other advantages are anticipated from the deliberate, coordinated application of ICT to public administration and political decision-making. Rapid outcomes, on the other hand, can only be expected if a solid institutional framework as well as suitable technical and infrastructure are already in place. e-Government will mostly be deployed in industrialized and advanced developing countries shortly [5]. Poorer countries, on the other hand, are seeing new opportunities. In many cases, the biggest impediment to transformation is not due to financial or infrastructure issues, but political stumbling blocks. e-Government can help partner countries devise and implement political and administrative reforms, as well as improve market-oriented frameworks, through development cooperation. In addition to the obvious advantages of new technology, e-government should be used to advance good governance and strengthen reform-minded players in politics and civil society.

1.1.3 Modernization Effect on Service Delivery

Shouzhi Xia [6] in his study on e-governance and political modernisation, elucidated that pro-poor basic delivery has the potential to increase stakeholder engagement in local public policy discussions. In line with the preceding, this paper concludes that overcoming the barriers to e-governance adoption and implementation is critical to achieving better public service delivery. ICT is frequently utilized with the purpose to boost public administration's efficiency and stated regulation. Allocated efficiency is a measure of how well service or infrastructure packages match consumer expectations. This indicates that the overall allocation of components, in precise, not only the allocation of factors by the public institution alone must be considered [7]. Even if higher public-sector service delivery costs high, allocated efficiency may be viewed as significantly higher if ICT-based solutions enable the issuance of a personal identity (Newman et al. 2008). In this setting, e-government is associated with New Public Management (NPM) techniques, which rank the effects of administrative activity to determine excellence.

Sullivan [8] has discussed Modernisation, democratisation and community governance.citizens receive timely and corrupt-free public services is therefore crucial. A

Strategic and Tactical Systems Level Framework and Model of Effective Public Service Delivery are created through this research with the comparison of utilizing benchmark framework to experience hassle-free service from e-government. The benchmarking framework was built based on research evidence from literature reviews, worldwide experiences, and case study analysis.

Arendsen et al. [9] investigated ways governments can improve service delivery by switching from reactive to proactive and proactive to predictive service delivery. The study aims to developing a transition model for e-government from one-stop shops to no-stop shops in order to overcome the challenges and complaints by Converging zero-form service delivery They suggested to make it successful, the government requires competent software trained personnel in software development, operations, information security and cryptography to manage the complexity.

1.1.4 Modernisation and Sustainability of e-Government Services

Nair and Prasad [10] said that modernizing government enhances an efficient public service delivery, addressing the development constraints, and fostering wellbeing is all possible with good governance. This might aid in achieving the Sustainability goals of development in the year 2030, the use of ICT to assist service initiatives continues to be a major factor in bringing about this shift, particularly in light of pressures from the global market and increased competition on a global scale. It provides amazing potential to innovate, enhance, and improve working methods while also providing several benefits. The greatest surge in digitization, however, will only be truly futuristic until it will assist each and every one in every corner of the globe. In order to achieve sustainability at all levels, it is essential to establish the strategies by recognizing the procedures, processes, and outcomes [11].

1.1.5 Challenges of Modernising e-Government

Several issues are hindering the development and construction of e-government.

(1) **Issues with institutional mechanisms**: It does not have a consistent top design. The premise of the e-government system is weak. The terms and regulations must be enacted.

(2) **Organizational and management issues**: It lacks a cohesive system. Task overlap, self-regulation, and bullish kind of management are still prevalent.

(3) **The issue of Wastage of Resources**: There is a disparity between localities. It is still typical to see on-going construction, low infrastructure usage rates, and significant information resource waste.

(4) **Data integration issue**: Public resource development is scattered, and data aggregation is challenging. Data towers, information enclaves, and digital voids still remain serious issues.
(5) **Performance evaluation issues**: Performance evaluation is inadequate in the e-government system. Poor performance results from a focus on construction rather than application.

1.2 Objectives of the Study

- To develop a Conceptual model to identify the antecedents of modernization of e-government services for sustainable development in e-government services.
- To design algorithm to validate and register citizen data to access the e-government services.
- To analyze the modernization effect of e-government performance with the integration of state of the art technology towards sustainable e-government services, across the globe, through a case study approach.

1.3 Review of Literature

1.3.1 Information Quality

Information systems are described as the standard of information generated by a certain system by DeLone and McLean [12]. The system's output is what is referred to as the system's output when defining the quality of the information provided. Information quality is a multifaceted concept that encompasses a wide range of attributes and characteristics. Experts use a variety of assessment indicators to determine the quality and value of information; as a result, different aspects of information influence the value of information. According to Wang and Strong [13], four variables determine the quality of information: (1) Intrinsic, (2) Accessibility, (3) Contextual, and (4) Representational.

The context of information quality on e-government websites is taken into account the characteristics of information quality. Experts and other organizations routinely offer definitions of e-government; nevertheless, only the World Bank [14] is considered in this study, which states: Government agencies employ information technology to transform connections with individuals, corporations, and other government agencies, is referred as e-government.

1.3.2   System Quality

E-Governance system quality refers to the performance and usability of the system among rural citizens. The operation quality of an e-government system is referred to as system quality, and it is a level that encompasses a variety of factors such as credibility, flexibility and response time [15]. The acceptance and success are influenced by the aspirations and interests of various stakeholders (Osman et al. 2014). Few studies have examined how users view these services, and the majority of these models are intended to support policymakers and practitioners in assessing the e-government services (Wang et al. 2005).

Because online service no longer requires face-to-face interaction and because the information system has replaced the window for online government services, the quality of e-government systems has become the users' first impression (Rai et al. 2002). System quality is therefore one of the most crucial metrics for assessing overall system performance and user perception.

1.3.3   Service Quality

The quality of e-governance services delivered to citizens is measured by service quality. Service Quality is defined by Zeithaml and Bitner [16], as Customer perception of the service provided based on responsiveness, reliability, empathy, assurance, and tangibility. This report focused on interactive and "one-stop" services while researching service quality. A mature electronic government service ought to be participatory in both directions. Personal taxes, license updates, fines, birth, marriage, and death announcements, among other things; enterprise registration, tax applications, customs, foreign exchange reports, and patent filing. e-government services integrate internal government systems with external internet interfaces to enable user interaction. In addition, an e-government system entails offering more extensive services for individuals, enabling them to conduct business more effectively [17]. Bhuiyan [18] conducted a research to see how the government prioritizes e-government services through profound service quality. The government of Bangladesh is working hard to attain this aim with the aid of quality services.

1.3.4   State-of-the-Art Technology

With the Case study approach, based on various model, it is suggested that it is a process consisting of 4 stages: strategizing, anchoring, enacting, and reinforcing. The stages identified and the constituent initiatives of each stage are all conceptual innovations, and in demonstrating the cyclical nature of the process the present study has revealed the path-dependent nature of Technology as well.

The failure to incorporate innovative and modern technology into e-governments is a major stumbling block to the deployment of e-government services. Frameworks and models must be updated regularly to keep up with technological advancements. It is critical to use cutting-edge technology for technological projects to last and remain competitive (e.g., e-governments). Governments should use more advanced and cutting-edge technology to promote e-government services because traditional technologies are becoming more complex and expensive to operate and maintain in developing countries. It is emphasized that cloud computing platforms should be used for e-government services [19].

Roland Traunmuller [20], a study on e-government was done, to modernize public administration and realign public governance. The e-government application landscape examined is divided into three categories: (1) promoting a knowledge-based economy, (2) providing residents and business customers with information and services, and (3) enhancing governmental cooperation.

1.3.5 Citizen Satisfaction

Van Ryzin [21] has defined Citizen Satisfaction as the combined decree of the citizens towards the quality performance of the local government. Gilmore and Souza [22] have stated that enhancing the responsiveness towards the mechanism of the public delivery system results in Citizen Satisfaction. To assess the performance of the services provided by the government, the level of service supplied to citizens must be assessed. Citizens are increasingly using the internet to interact with different government services, raising the standard for efficient service delivery. There is a greater emphasis on enhancing citizen happiness and loyalty to government e-services, which result increasing of the public participation. Even though, Kuwait has over 3,000,000 internet users, the percentage of citizens who are satisfied with e-services is not high. This shows Government needs to focus more on citizen satisfaction than concentrating infrastructure.

Sayed et al. [23] conducted a study to determine the maturity of the organization. The findings of the study shows that Kuwait intends to improve citizen services to promote citizen happiness. This would be impossible to achieve without first assessing present services and developing KPIs. The major goal is to implement an e-government solution in Kuwait, particularly in the educational sector, in order to make decision-making simpler and more efficient. This research lays forth a strategy for implementing KPIs for e-government services. The proposed technique constructs and measures key performance indicators based on mission, vision, and objectives (KPIs). The author has employed five major measures: loyalty, involvement, productivity, communication, and satisfaction.

This has a big impact since, despite having good ICT infrastructure; the proposed road map emphasizes the need to strengthen e-government services such as increased teacher and student training, the construction of new schools, the design of long-term educational policies and plans for Kuwaiti residents to deal with significant advancements in the ICT industry with all feasible improvements.

1.3.6 Efficiency of e-Government Services

The efficiency of e-government services determine citizen satisfaction. One of the major compelling arguments for modernizing e-government reform is that it improves government institutions' internal or production efficiency, to save the money of people who are paying taxes. There are two methods to get done with this theory. One instance is the automation of application management and process simplification to increase productivity for reducing employment. Another goal is to lower public procurement costs by boosting price transparency, constantly promoting competitiveness in the global market, and instituting more accessible and market-friendly acquiring procedures. These savings, however, may be counterpoised by costs that limit or eliminate such adeptness expansions, as discussed further below. The integration of the advancement of technology in the present system improves efficiency. e-Government is viewed as a tool for,

1. Increasing the efficacy of civic government,
2. Improvement in the public service delivery, and
3. Strengthening the ingenuousness and clearness of political processes.

   Shouzhi Xia (2017) conducted a study to investigate the factors that can help to enable e-governance. Government transparency, offline e-government services engagement, and the level of naivety are the three components of political modernization. After evaluating secondary data from numerous databases, the researchers concluded the following things. To begin with, greater political modernization in Asia will start of expansion of e-governance. The availability of open data, in particular, improves government openness. Offline political participation and liberty rise as a result of e-participation. Second, because open of data and electronic participation influence various areas of political modernization, deciding which characteristic affects the process of e-governance and makes the modernization challenging. The findings suggest that the government must highlight the role of e-governance in order to enhance political modernization.

1.3.7 Sustainable e-Government Services

World commission on environment 1987 explains that the terms, Sustainability and sustainable development are often used interchangeably to denote developing pathways that meet present needs to adhere to the capacity of to fulfill the requirements of upcoming generations. The ability of a government to implement e-government services effectively while addressing the needs of many stakeholders is referred to as sustainability. Sustainability focuses on two main characteristics of e-government service implementation and citizen adoption. Citizens-centric, trustworthy e-government services that leverage cutting-edge technology to provide an efficient, effective, cost-effective service, as well as full participation and satisfaction from all users, define sustainable electronic government services.

A study on the effects of electronic government on several aspects of growth concerning sustainability in the Middle East and North African countries area was published in 2021 by Iyad Dhaoui. This research represents data from more than fifteen countries. The preliminary part of this study includes the role of electronic government services in good governance. The next section describes the effect of good governance on long term development and the impact of development of electronic services provided by the government. The results demonstrate that the majority of good governance practices promote sustainable or long-term development. Research reveals that digitalization improves regulatory quality to a lesser extent and boosts government effectiveness and corruption control when it comes to the impact of development on governance parameters. Rather than being a catalyst for progress, e-government-related variables affect several areas of sustainable development, contrary to expectations. The study makes some recommendations based on these findings. Fundamentally, regulations governing the use of digital technology must be properly integrated into public sector reform. Middle Eastern and North African (MENA) nations should enhance their institutions for education and skill development, boost accountability, and improve business conditions.

1.4   Proposed Conceptual Model

The researchers have proposed a model to integrate the State-of-art-of technology in the name of modernization to attain sustainability in e-government services (Fig. 1.2).

The Conceptual Model is developed based on the systematic literature review that is based on DeLone and McLean's Information System Success Model developed in the year 2003 by Joshi and Islam [19]. The proposed conceptual model explains how information quality and system quality contribute to the service quality of electronic government services.

The proposed model validates Service delivery and improves the efficiency of e-government services and citizen satisfaction and involvement in participating in the e-government services. The proposed conceptual model is framed to attain sustainable

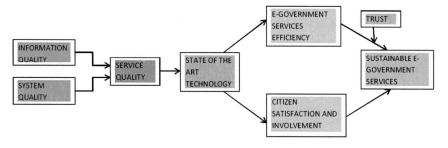

Fig. 1.2 Author's conceptual model

e-government services through state of the art technology. This can also be achieved efficiently through improving service delivery and citizen satisfaction. The model also analyses the impact of state-of-art technology, i.e., artificial intelligence and advancement of technology results.

1.5 Methodology

The study is based on secondary data collected through systematic literature review. The methodology followed is the combination of development of conceptual model, designing of data algorithm and case-study approach. Various factors responsible for successful implementation of e-governance services followed across the globe are scrutinised to develop a conceptual model. Simple data algorithm is developed for validation and registering data from the citizen. Under case study method, best e-governance services, modernisation towards sustainable e-government services are analysed in detail.

1.6 Algorithm

1.6.1 Algorithm for E-Filing a Form

The purpose of the development of the algorithm is to simplify the validation and registration of citizen data. This was done in the following steps in the flow chart.

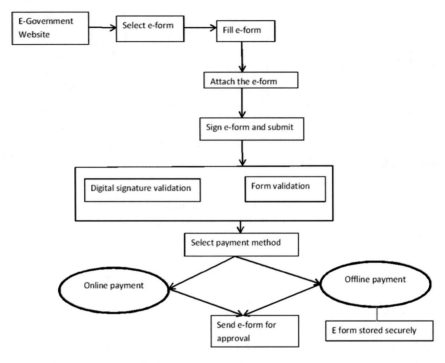

The above is an example for an algorithm in a particular e-government service to fill an electronic bank form.

Step 1: Go to the e-government website.

Step 2: Select the e-form which is specific to be filled.

Step 3: Fill in all the particulars requested in the e-form with the appropriate data.

Step 4: Attach the e-form to the required page of the bank website.

Step 5: Sign the e-form with the original signature and then submit.

Step 6: Validate the signature with the already obtained data available in the bank data and validate the particulars in the e-form whether all information provided is true.

Step 7: Select the method of payment that is accessible to pay the money, whether to be paid online or offline.

Step 8: Now send the e-form for approval.

Step 9: Until the bank accepts the e-form, the data of the customer will be saved securely in the government database with due respect to privacy.

Note: Each e-form or application will contain a separate Serial Request Number (SRN) through which the application can be identified.

1.6.2 Algorithm-2

The below is the algorithm to Register and validate the personal data on a particular e-government service.

Step 1: Start the process by clicking the start button provided on the website.

Step 2: Enquire the member about the registration process by asking for the identity number or registration number.

Step 3: Validate the email id of the user, if the email id provided is incorrect or not stored in the database, end the process.

Step 4: If the email address provided matches the Name in accordance with the web community's religion, verify the computer's linguistic reference.

Step 5: If i-Name confirms the regulations, now search the information track of the web personality.

Step 6: Now verify the personal data of the web member, and if no information is found, put it on the mark for permanent monitoring.

Step 7: If the information is found to be valid, allow for the registration and proceed to the post-registration analysis of the web member.

Step 8: If negative data is found, registration of the member will be failed.

Step 9: If the data of the member is valid, add the member to the list of reliable members.

Step 10: If the data is found to be not reliable, send a notification to change that specific data.

Step 11: Add that member to the list of unreliable members.

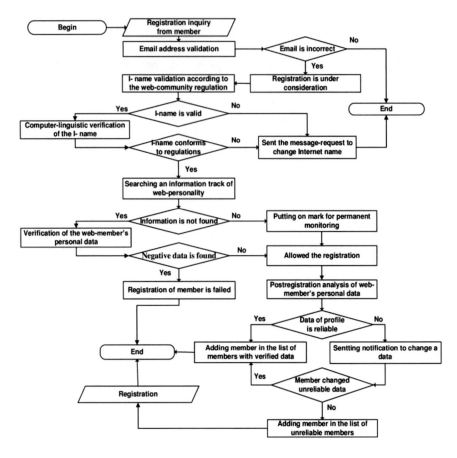

1.7   e-Governance Initiatives Across the Globe—Case Study

1.7.1   Denmark

E-Governance helps to boost the public administration efficiency, enhance the delivery of electronic services, and promoting the transparency of government processes from the standpoint of all factors in the reform process. This is also one of the reasons why the European Commission, named Denmark the most digitalized country in the world. Furthermore, Denmark was ranked first in the world in the UN's 'e-Government Survey 2018'. When it comes to public digitization, no country performs better than Denmark among the 193 member countries. The highlighted feature of Denmark's e-government services is that every Dane has a secure digital key known as a NemID. NemID is a single key for all of Denmark's digital transactions.

1.7.2 Jordan

Jordan has taken several initiatives to strengthen ICT infrastructure as a precondition for e-government adoption, capitalizing on the golden opportunity afforded by its people resources, which can be a useful asset in aiding the growth of ICT projects in Jordan. The unique activities for modernizing the e-government in Jordan are as follows:

- To provide the required information with high accuracy on time.
- To provide required infrastructure and technology to increase the level of security of information.

In the year 2000, Jordan's Ministry of Information and Communication Technology launched an e-government initiative [24]. The goal of the initiative is to provide Jordanians with access to government information and services regardless of their location, education, economic status, or IT skills [25].

Jordan's government has been working to expand the number of e-government services available to change how citizens interact with and participate in government. Nearly 150 e-government services are now available in all government institutions. Although the Jordanian e-government policy has been in place for more than eighteen years, it has failed to enhance citizen contact with e-government services [26, 27]. This is supported by UN studies, which reveal that Jordan's e-government development score has raised up from 50 in 2010 to 98 in 2018 [28, 29].

1.7.3 Bangladesh

In Bangladesh, electronic birth registration was successfully implemented by the municipal administration in 2001. The UNICEF-funded project required a $20,000 investment and operational costs per month of approximately $200. Consultations and data management now will just take a fraction of the time they used to initially. Manually gathered data transfer errors can now be avoided. e-Government offers poor countries a tremendous opportunity to improve their administration and governance. Bangladesh is eager to take advantage of the situation. By studying associated policies and legislation and analyzing numerous programs, the authors discovered that the country achieved tremendous progress in delivering public services. Bangladesh is extending e-government to every feasible sector, including education, agriculture, poverty alleviation, health, trade and commerce, and so on.

e-Government offers poor countries a significant chance to improve administration and Government Bangladesh is eager to take advantage of the opportunity. Bangladesh is extending e-governance to every feasible sector, including education, agriculture, poverty alleviation, health, trade and commerce, and so on, in order to improve the transparency of governmental services.

1.7.4 Hong Kong

Hong Kong is ranked number 18 in the global ranking of the e-government index. For the purpose of delivering a better service quality to the citizens the Hong Kong Government is proactively entertaining modernized electronic services. Significant aspects of e-government services in Hong Kong is the one-stop web portal which helps to provide quick and ease of access of the information to the people and businesses. About 75% of the citizens of Hong Kong are well known about the e-government services provided by their nation, and among these 75% nearly 70% of the citizens are using them.

The level of maturity of the government of Hong Kong in the field of IT and e-governance has made it a pioneer to look from citizens' perspectives and the level of citizen satisfaction with the e-government services.

Citizens can make use of E-TAX to file through online services, their details of income tax returns, and OABS for scheduling appointments with all kinds of government offices and E-PORTAL helps to deliver important government information and also serves as a portal to websites specifically government agency, to provide more in-depth information. These three services are illustrative, educational and transactional services with varying levels of complexity, demonstrating the generalizability of findings.

1.7.5 China

Countries in Asia have noted the importance of developing a perfect e-government system in the last ten years. Other fields, including offline political participation, are being impacted by the development of e-government. Citizens' political participation, according to Jiang and Xu [30], may have unforeseen repercussions as a result of China's subtle reform of e-governance. He and his colleagues (2017) revealed that a growing number of ordinary Chinese citizens are engaged in environmental policy-making processes.

e-Governance appears to have a clear impact on citizens' liberty. For starters, once the e-government system is fully operational, citizens will have more access to official data. Citizens previously did not have this level of liberty because even if they requested it legitimately, they couldn't get the information they needed from the government. People can now obtain the information they need from a government website or database.

They are free to request additional information via internet platforms. Then, citizens have more options when it comes to accessing government services because they can pick between e-government services and offline transactions. This circumstance beyond people's vision was beyond the development of e-governance. In other words, citizens were required to go through a series of complex and time-consuming procedures while dealing with potentially rude government employees. Using online systems, a lot of commerce can now be done online and have an offline effect.

1.7.6 Kuwait

This study presented a road map for measuring KPIs based on data mining, sentiment analysis, and statistical methods. The proposed road map is dominated by Kuwait's e-government performance. The KPIs measurements are investigated by using a questionnaire to collect data based on data mining, sentiment analysis, and statistical methods, this study presented a road map for measuring KPIs. Kuwait's e-government performance dominates the proposed road map. The KPIs are measured using a questionnaire, which is collected and reviewed throughout the process using Google forms and traditional methods. In addition to a sentiment model with a logistic regression classifier, the participants' feedback was used to cluster the questionnaire output using the Weka data mining tool. This is to use Google forms and traditional methods to apply classification based on the feedback received and to review it throughout the process. In addition to a sentiment model with a logistic regression classifier, the participants' feedback was used to cluster the questionnaire output using the Weka data mining tool.

Statistical methods such as mean and standard deviation are also used. According to the findings, despite substantial investment in the ICT sector, the impact is insufficient to provide citizens and users with a high level of satisfaction. Data collected from the government's dashboard should be used in future information science research on Kuwait's e-government. In addition to improving the management approach, scientists can use fuzzy logic methods to increase citizen satisfaction.

The following table portraying the e-government development index of top-performing countries around the globe in the field of e-governance. Source: Government website (Tables 1.1 and 1.2).

1.8 Analysis and Discussions

Denmark is the pioneer in e-government services. The researchers have studied the success factor of every country to implement state of the art technology. The algorithm developed is just a sample for validating and registering the data of citizen. Many such algorithms can be developed based on the requirement of enhanced services.

The researchers have framed the conceptual model as the foundation of the study which explains the antecedents of the modernisation of e-government for achieving sustainable e-government services through state of the art technology. The case study of six countries which are providing high service quality with the integration of state of the art technology was reviewed and discussed to find out the success factor behind sustainable e-government services. The researchers could justify that the

Table 1.1 e-Government development index

Country (Characteristic)	Index score
Denmark	0.98
Republic of Korea	0.96
Estonia	0.95
Finland	0.95
Australia	0.94
Sweden	0.94
United Kingdom	0.94
New Zealand	0.93
United States	0.93
Netherlands	0.92
Singapore	0.92
Iceland	0.91
Norway	0.91
Japan	0.9

Table 1.2 Evaluation and comparison of e-government performance index

	Denmark	Bangladesh	Jordan	China	Hong Kong	Kuwait
Information quality	High	High	Medium	High	Low	High
System quality	High	High	High	High	High	High
Service quality	High	Medium	Low	High	Low	High
State of the art technology	High	Medium	Medium	High	Medium	High
e-Government service efficiency	High	Medium	High	Medium	Medium	High
Citizen satisfaction and involvement	High	High	High	Medium	Medium	Medium
Sustainable e-government services	High	Medium	Medium	Medium	Very low	High

Source Website stata

variables used in the conceptual model namely Service Quality, System Quality and Information Quality provide a support for the state of the art technology which is practiced in the countries discussed in our case study. Service delivery is a component of service quality, which is one of the main reasons of Sustainable e-government services in Hong Kong and Bangladesh. The main reason for successful e-government services in China is evidenced that addressing the needs of the citizen especially the person with disabilities forms the part of Information Quality. The practice of providing renowned System Quality has influenced the performance of sustainable e-government services in Kuwait which is a proof for state of the art technology.

According to Malodia and Mishra [31], the proposed conceptual model explained the effect of system quality and information quality on the effectiveness of service delivery.

Citizens' trust in technology has an impact on their desire to use the services online. However, neither the issue of trust nor its impact on individuals' desire to use online public services has been extensively investigated to date. The value of addressing cultural features in online trust research has already been acknowledged in literature in similar contexts. This report presents the findings of a comprehensive literature evaluation of existing trust research. The author concludes that, in spite of its crucial importance, no parallel study has been undertaken that reveals individuals' decision-making mechanisms for adopting e-government services, with a particular focus on the issue of trust.

Although some developing countries will be able to fill the governance breach with affluent countries, the new ICT is projected to favour the countries that are already in a stronger position, adding to the global governance split.

e-Government can drastically differentiate the internal and external functions. It is a tool for reforming or transforming not only the existing operations but also the way, government does business and delivers services. The government should create intuitive interface that is simple to use. The IT employees' job is to design simple predesign simple process by the service providers, so that citizen may complete transactions with very less human interaction.

Government should communicate information through many channels and instruct how to use the services. The government and IT people should ensure that up-to-date security protocols are used. The government should also frame appropriate private policies to ensure that the personal data has not been exploited.

Reforms inside specific administrations, according to the findings, are typically just the first step toward better labour productivity through e-government. The pressures on ICT-based agency collaboration are increasing as multilayer systems get more complex. When ICT is introduced into government, it kick-started modernization processes that lead to more results-driven and customer-focused management. Instead, it seems that the creation and implementation of e-government initiatives speeds up the process when governments start administrative modernisation. e-Government improvements, on the whole, give citizens more freedom. They felt that by living in this manner, they would foster "good governance".

The level of citizen satisfaction is positively influenced by the quality of service offered, which further promotes the modernization of e-government, aided by data and algorithms.

The modernization of e-government in five nations was examined through a case study approach. The major highlights of the e-government services of those countries are discussed below.

- Trust is the most important factor of success for e-governance in Denmark.
- Hong Kong's e-Government Strategy increases the efficiency of service delivery while simultaneously improving the quality of life of its inhabitants.

- Progress in online service delivery and innovative technology solutions have gained special recognition in Bangladesh.
- Addressing the requirements of people with disabilities, Beijing has worked tirelessly to strengthen social security and public service systems by pushing new online apps. This has been a crucial aspect of China's success.
- Fuzzy logic methods used by the government has been considered a strong support behind the success of e-government service in Kuwait.

1.9 Conclusion

The use of big data and algorithm in the process of modernization of e-government results in a reduction in the time and cost of the services for the citizen, the number of duplicated government functions, and an improvement in the quality of government services. There is also reduction in the time it takes to service applicants, a reduction in administrative barriers, an expansion of the range of public services, an improvement in the efficiency of government agencies, an increase in customer satisfaction, and an overall improvement in public governance. This study concludes that the information society is evolving rapidly, and significant measures have been done to mould and incorporate e-government into the current public governance system. It will swiftly expand its efficiency and costs will be fully justified with consistent funding and proper control, as has happened in the top Western countries. From the previous studies it's evidenced that, there is a slow growth but quick shift toward development, modernization, and increased efficiency and effectiveness in the public government system. It's also worth noting that the installation of modernized e-government plays a critical part in this process.

References

1. Bhattacharya, D., Gulla, U., Gupta, M. P. (2012). Emerald article: E-service quality model for Indian government portals: citizens' perspective. *Journal of Enterprise Information Management.*
2. Morgeson, F.V., Mithas, S. (2009). Does E-government measure up to E-Business? Comparing end user perceptions of US federal government and E-business web sites. *Public Administration Review.*
3. Iqbal, M., & Seo, J. W. (2008). e-Governance as an anti-corruption tool: Korean cases.
4. Ziyadin, S., et al. (2020). Digital modernization of the system of public administration: prerogatives and barriers.
5. Newman, J. (2001). Sage publications, *Modernising Governance.*
6. Xia, S. (2017). E-governance and political modernization: An empirical study based on Asia from 2003 to 2014. *Administrative Sciences.*
7. Saxena, K. B. C. (2005). Towards excellence in e-governance. *International Journal of Public Sector Management. 18*, 498–513.
8. O'Sullivan, M. (2001). The political economy of comparative corporate governance. *Review of International Political Economy, 10*(1), 23–72.

9. Arendsen, R., Peters, P., Ter Hedde, M., & van Dijk, J. (2014). Does e-government reduce the administrative burden of businesses? An assessment of business-to-government systems usage in the Netherlands. *Government Information Quarterly, 31*(1), 160–169. ISSN 0740-624X.
10. Nair, K. G. K., & Prasad, P. N. (2002). Development through information technology in developing countries: Experiences from an Indian State. *The Electronic Journal of Information.* Wiley Online Library.
11. Dhaoui, Iyad, (2019). Good governance for sustainable development, MPRA Paper 92544, University Library of Munich, Germany
12. DeLone, W. H., & McLean, E. R. (1992). Information systems success: The quest for the dependent variable. *Journal of Management Information Systems.*
13. Wang, et al. (2005). *Journal of Comparative Economics*, Empirical linkages between good governance and national well-being.
14. The World Bank Annual Report (2012). *1.* Main Report.
15. Delone, W., & McLean, E. (2003). The DeLone and McLean model of information systems success: A ten-year update. *Journal of Management Information System, 19*(4), 9–30.
16. Zeithaml, V. A. & Bitner, M. J. (2003). Services marketing: Integrating customer focus across the firm. 3rd Edition, Irwin McGraw-Hill, New York.
17. Chohan, S. R., Hu, G., Si, W. and Pasha, A.T. (2020). Synthesizing e-government maturity model: a public value paradigm towards digital Pakistan, *Transforming Government: People, Process and Policy, 14*(3), 495–522. https://doi.org/10.1108/TG-11-2019-0110
18. Bhuiyan, S. H. (2011). Modernizing Bangladesh public administration through e-governance: Benefits and challenges. *Government Information Quarterly.*
19. Joshi, P., & Islam, S. (2018). *e-Government Maturity Model for Sustainable e-Government Services from the Perspective of Developing Countries.*
20. Traunmüller, R., & Wimmer, M. (2004).E-Government: The challenges ahead. *International Conference on Electronic Government.*
21. Jilke, S., Van Ryzin, G. G., & Van de Walle, S. (2016). Responses to decline in marketized public services: An experimental evaluation of choice overload. *Journal of Public Administration Research and Theory, 26*(3), 421–432.
22. Gilmore, A., & D'Souza, C. (2006). Service excellence in e-governance issues: An Indian case study. *Joaag, 1*(1), 1–14.
23. Sayed, et al. (2019). *Journal of Information Technology Research*, Can gamification concepts work with E-Government?
24. Almarabeh, T. & Abu Ali, A. (2010). A general framework for E-Government: Definition Maturity Challenges, Opportunities, and Success. *European Journal of Scientific Research, 39*, 29–42.
25. Alrawabdeh, W. (2017). E-government diffusion in Jordan: Employees' perceptions toward electronic government in Jordan. *American Journal of Applied Sciences.*
26. Rana, N. P., & Dwivedi, Y. K. (2015). Citizen's adoption of an e-government system: Validating extended social cognitive theory (SCT). *Government Information Quarterly, 32*(2), 172–181.
27. Kanaan & Masa'deh (2018). A theoretical perspective on the relationship between leadership development, knowledge management capability, and firm performance, *Asian Journal of Social Science.*
28. United Nations (2010). Report of the Secretary General.
29. United Nations (2018). Report of the Secretary General.
30. Jiang, X., & Li, Y. (2009). An empirical investigation of knowledge management and innovative performance: The case of alliances.
31. Malodia, S., Dhir, A., Mishra, M., & Bhatti, Z. A. (2021). Future of e-Government: An integrated conceptual framework. *Technological Forecasting and Social Change.*

Chapter 2
e-School Initiatives that Instigated Digital Transformation in Education: A Case Study According to SABER-ICT Framework

Tahani I. Aldosemani

Abstract e-Government represents a fundamental shift in the way governments around the world are embracing their missions. The waves of e-government are growing through public organizations and administrations worldwide using Information and Communication Technology ICT to provide services between government agencies, citizens, and businesses. Regular and sufficient use of ICT in schools can enhance e-government endeavors. The enhancement of digital skills related to e-government starts in schools; therefore, ICT implementation in education will simultaneously reinforce e-government optimization and enhance education quality. Investments in ICT in education can make education systems more resilient to future shocks and help reform and reimagine how education is delivered. Educational ICT can support governments build a more efficient business case for investing in unserved areas, enabling effective financial models for better distribution of funds while supporting efforts to extend broadband connectivity to every school. Effective ICT implementation in education will inform programs across different sectors and extend the reach of schools to a more significant number of students and teachers. In this case study analysis, ICT implementation in education is discussed according to the *Systems Approach for Better Education Results* (SABER)-ICT framework focusing on Saudi Arabia's efforts. The framework is intended as a tool for analysis and guidance to benchmark the current state of ICT/education policies in countries and shape future directions and initiatives. The Saudi case is discussed according to the framework's eight main domains shedding light on critical outcomes achieved within each domain. The case study is concluded with several recommendations and implications for practitioners.

Keywords e-Government · ICT in education · Digital transformation · e-Schools · SABER-ICT framework

T. I. Aldosemani (✉)
Prince Sattam Bin Abdulaziz University, Alkharj, Saudi Arabia
e-mail: t.aldosemani@psau.edu.sa

2.1 Introduction

The world is amidst a technological revolution, and learners must be adequately prepared to thrive in this rapidly changing world. Governments, international organizations, industry, and civil society must collaborate and coordinate efforts to bridge the gaps in digital and literacy skills among learners. The United Nations Educational Scientific and Cultural Organization (UNESCO) [1] report highlighted that 3.6 billion people still have no access to the Internet, and about 258 million children are out of school [1]. The UNESCO [1] report, The Digital Transformation of Education: connecting schools, empowering learners, discusses core principles to support governments in adopting more comprehensive school connectivity plans proposing a framework to leverage school connectivity based on four pillars: Map, Connect, Finance, and Empower. According to the report, articles 28 and 29 of the United Nations Convention on the Rights of the Child (UNCRC), all children must have access to high-quality education. Connectivity and access to information can serve as a means to empower schools and the communities surrounding them beyond educational goals. For example, the Broadband Commission for Sustainable Development advocates the power of ICT and broadband-based technologies for achieving Sustainable Development Goals (SDGs) [1, 2]. These initiatives prioritize the development of broadband infrastructure and services to raise global community awareness through a multi-stakeholder approach, particularly in developing countries and underserved communities.

The ITU/UNESCO report discusses the outcomes of broadband, digital transformation policies, and regulatory frameworks on the development of the digital economy, emphasizing through evidence the significance of regulatory and institutional variables in enhancing digital transformation plans. In addition, the report highlights the impact of supported broadband technologies and effective Information and Communication Technology ICT regulations on the development of the digital ecosystem and national economies. The Commission report set seven targets; four of them focus on affordability, connectivity, digital skills, and empowerment of youth and adults. Developments in ICT reinforce governments' initiatives for improving the education system by providing data regarding the number of children out of school, developing a better understanding of factors that impact learning outcomes, and ensuring that government services reach everyone and, more specifically, the most vulnerable groups. Indeed, connectivity, access, and obtaining accurate data about schools can promote the achievement of three goals of the SDGs: promoting high-quality education and lifelong learning (SDG4) and equal access to opportunity (SDG10) and eventually reducing poverty (SDG1) [1].

Educational ICT can support governments build a more efficient business case for investing in unserved areas, enabling effective financial models for better distribution of funds while supporting efforts to extend broadband connectivity to every school. In addition, effective ICT implementation in schools will inform programs across different sectors for better resource allocation, coordinate delivery and response

during national or global emergency crises, and enhance mapping of the vulnera-
bility of the communities surrounding schools [1]. Indeed, schools are at the center
of most societies, and empowering schools with adequate ICT can help transform
these education facilities into digital hubs that could spread innovation to the whole
community. Digital education transformation through ICT fundamentally requires a
comprehensive multi-phased government plan that starts with mapping schools for
connectivity, financing school connectivity, and empowering learners [1, 3]. World
Economic Forum [3] discusses multiple international initiatives in adapting for innovation in connectivity-enabled learning and teaching models in which infrastructure can be deployed to provide student connectivity. The case study report recommends revolutionizing schools' connectivity to future-oriented education connectivity programs and reaching 24/7 connectivity for students in and out of school,
leveraging technologies that can help achieve such goals through cost efficiency and
scalability.

Strategic leadership and governance during the implementation of ICT supported
by reliable data will facilitate a better evaluation of existing digital inequalities in
schools. Such data will provide policymakers with effective feedback on future policy
and action changes. In this regard, e-government is a powerful approach to support
different types of economies to address the digital divide challenges in education,
bring the benefits of the emerging global information society to their respective
populations, and make it more responsive to their needs. Saudi Arabia has made
substantial efforts to enhance digital inclusion for all to accelerate its progress toward
achieving the United Nations Agenda 2030, SDGs (UN 2018), and Saudi Vision
2030 [4]. Saudi Arabia developed special initiatives and implemented systematic
measures to ensure meaningful connectivity and access to e-government services.
Among these initiatives are the focus on digital literacy and skills for all, including
vulnerable groups.

2.1.1 e-Government

E-Government represents a fundamental shift in how governments worldwide
embrace their mission [5]. Baum and Di Maio [6] define e-government as the "continuous optimization of service delivery, constituency participation and governance by
transforming internal and external relationships through technology." e-Government
can also be defined as using Information and Communications Technologies (ICT)
to transform government processes by making them more accessible, effective,
and accountable [7]. e-Government supports cost-effectiveness in government and
general operations, improves transparency and accountability in public decisions, and
strengthens continuous contact with citizens regardless of geographical location or
population level. Further, it enhances the emergence and sustainability of e-cultures.
e-Government enables greater access to information, promotes civic engagement
and interaction, and reinforces development opportunities for rural and traditionally underserved communities [8]. E-government has three key goals; using ICT to

expand access to government information, broadening civic participation in government, and making government services available online [7]. Therefore, e-government focuses on using ICT to transform the structures, operations, and, most importantly, the government culture among all sectors, including education.

E-government is an essential tool for overall reform that transforms relations with citizens, the private sector, and government agencies, promoting citizen empowerment and improving government efficiency [9]. E-government functions can be classified into four main categories: Government-to-citizens (G2C), providing citizens and others with comprehensive electronic resources to respond to individuals regarding government transactions with continuous communication between the two, and increasing citizens' participation. Government to business (G2B) provides various service exchanges between the government and the business sectors, including distributing policies, rules, and regulations. Government-to-government (G2G) refers to the online communications between government organizations, departments, and agencies based on a super-government database and the relationship between the government and its employees. Government-to-employee refers to the relationship between the government and its employees to serve employees throughout managerial duties and processes. Seventeen challenges and opportunities of e-government implementation have been discussed, including infrastructure development, law and public policy, digital divide (e.g., e-literacy and accessibility), trust, privacy and security, transparency, interoperability, records management, permanent availability and preservation, education and marketing, public/private competition and collaboration, workforce issues, cost structures, and benchmarking [10].

In Twizeyimana and Andersson [11], the authors conducted literature review research on the public value of e-government to explore the current state and the added value it is supposed to yield by adopting Ndou's e-government framework (2004). The value creation potential needs to be understood as a constituent of the success of e-government projects. According to the literature analyzed, understanding public sector values is key to designing e-government projects and creating a successful alignment among different actors in the public sector. Further, the success of e-government can be a result of citizens' perceptions of the embedded value of using those systems. The understanding of e-government and the value it yields is linked to the public sector management and stakeholder groups and their interests' encompassing citizens, businesses, governments, and employees and their interrelationships.

Further, the study discusses e-government application domains of e-Administration, e-Citizens and e-Services, and e-Society. The study identified these three main dimensions of the public value of e-government and their related six dimensions through the analysis of 53 articles. These six dimensions include improved public services, administrative efficiency, open government capabilities, ethical behaviors and professionalism, social value and well-being, and trust and confidence in government. The public value of e-government is analyzed from a transformation perspective, internal, external, and relational areas, as well as from the standpoint of users, stakeholders, and their interrelationships. The findings revealed a lack of research on the value of e-government in many countries in general and developing countries in specific.

Further, the dimension of open government capabilities is predominant in all e-government domains and dimensions in the conceptualization of e-government. The study concludes that public expectations of e-government are generally based on the ultimate goal of improving the relationships between the citizens and the state. Finally, the study proposes a descriptive framework that can enhance the understanding of the public value of e-government from different dimensions and highlight the interrelatedness between these dimensions to support the government in its endeavors to assess the performance of e-government initiatives.

Malodia et al. [12] identify the digital divide, economic growth, political stability, perceived privacy, and shared understanding as moderating conditions of e-government. The study defines e-government as a multidimensional construct with customer orientation, channel orientation, and technology orientation as essential processes that need to precede e-government initiatives. Two key goals guided the study, first, to conceptualize and define e-government as a multidimensional construct, and second to identify the factors that should precede e-government initiatives across various disciplines with the analysis of potential consequences and moderating variables. The study proposes a conceptual framework for e-government based on three subcategories; empowered citizenship, hyper-integrated networks, and evolutionary systems architecture to ensure efficient delivery of government services with transparency, reliability, and accountability. Malodia et al. [12] identified three necessary stages to ensure citizens' empowerment in e-government. These stages include ensuring citizens' inclusivity, maintaining free availability of information, and empowering citizens to participate and influence the decision-making process.

Meiyanti et al. [13] discussed e-government implementation challenges in developing courtiers by analyzing 18 studies highlighting e-government challenges. The Preferred Reporting Items for systematic reviews adopted in this study show that challenges facing e-government can be categorized into five key domains: Information Technology (IT). infrastructure, managerial issues, digital culture, laws, and legislation, and budgeting are influenced by different conditions, including social, economic, political, cultural, education. Within the (IT) infrastructure, the literature identified a lack of technological skills among leaders, employees, citizens, and disabled people and a lack of qualifications required of government (IT) staff and developers. Further, and among the (IT) infrastructure domain, is the lack of hardware and software and maintenance processes, lack of communication systems, digitized information, integration systems, interoperability, record mobility, and lack of data availability and preservation. Further, among the managerial challenges are the lack of proper top-down administrative processes, resistance to change, lack of transparency, lack of coordination and collaboration, and workforce turnover. In addition, among laws and legislations are the challenges of lack of automation efforts for documentation processes and transactions and lack of policies that support e-government endeavors.

Regarding digital culture, the literature identified the digital divide, lack of awareness, and lack of trust. Finally, budgeting represents another critical challenge including lack of budgeting, financial resources, corruption, and lack of proper management of available resources. These challenges can serve as a starting point to stimulate discussion on mitigation strategies and how to improve internal and external e-government efforts.

Indeed, countries with resilient and high-quality education systems are usually known to provide sufficient technological knowledge and advanced digital skills to their citizens to use a wide range of digital applications and other technical devices and programs such as e-systems and e-services [8, 14]. Further, consistent and reliable Internet connection in schools is essential to successfully transform education from traditional to digitally mediated learning. Providing students with an immense corpus of information is often considered a significant indicator of quality education [15]. Therefore, regular and sufficient use of ICT in schools can enhance e-government endeavors by mitigating the digital divide caused by a lack of e-literacy skills or access challenges. The enhancement of digital skills related to e-government starts in schools, therefore, ICT implementation in education will simultaneously reinforce e-government optimization and enhance education quality. Furthermore, according to Chen et al. [15], increased spending on internet service in schools can lead to improvements in eight academic performance indicators, including three college readiness indicators in high schools (e.g., Scholastic Assessment Test (SAT)/American College Test (ACT) meet criterion rate) and five commended performance indicators on grades 3–11 (e.g., math, reading). Increased funding for ICT can alleviate access to the Internet and reduce the digital skills divide, which represents one of the main challenges to successfully implementing e-government initiatives and innovation plans in general [16]. To summarize, improved education quality in schools and enhanced level of digital knowledge and skills, including skills related to e-government and other digital service platforms, significantly rely on ICT implementation in schools [5, 17].

2.1.2 ICT-Enhanced Schools

ICT in schools has a significant role in enhancing and optimizing information delivery. ICT-enhanced schools or e-schools use ICT to improve learners' knowledge, skills, and attitudes [15]. These schools support qualified and competent school management teams and teachers who use ICTs in their day-to-day practices [18]. Solid, resilient, and cost-effective infrastructure solutions can help ensure a smooth rollout of school educational technology initiatives [19]. Further, investing in Cloud computing enables schools and districts to adapt immediately in times of crisis and maintain education continuity at scale [20].

The UNESCO ICT Competency Framework for Teachers (ICT-CFT) serves as an asset for countries to develop holistic national teacher ICT competency policies and standards that can be integrated into ICT in education plans. The framework emphasizes how technology can support understanding ICT in education curriculum and assessment, pedagogy, application of digital skills, organization and adminis-tration, and teacher professional learning. Further, available literature suggests that ICT can facilitate students' achievement of learning outcomes and lead to better teaching methods. The increased use of ICT in education can significantly impact students' knowledge, presentation skills, and innovation capabilities while encour-aging students to put more effort into their learning. Further, ICT increases the reach of institutions to a more significant number of students and facilitates faculties' training because of the off-the-classroom learning technologies and resources such as Massive Open Online Courses (MOOCs) [5, 15].

In addition, ICT can help overcome the physical constraints of classrooms, leverage mobility, and facilitate access to course materials through mobile devices using digital repositories for lectures, course materials, and digital libraries. Further, ICT can potentially enhance cloud-based educational management systems for effi-cient and sustainable educational resources [21, 22]. ICT in schools enables the application of the flipped classroom approach utilizing portable devices. It can also help prepare students to compete globally and be active and skilled participants in the workforce by providing opportunities to learn new skills through facilitated access to online learning resources. Furthermore, ICT minimizes the cost and time associ-ated with information delivery and automates regular day-to-day school tasks. More importantly, ICT supports the administrative processes of institutions to enhance the quality and efficiency of their services, backed by flexible assessments and reporting tools that can facilitate data-driven decision-making for an easy-to-manage learning environment.

Two major and recent global initiatives aimed at leveraging ICT in schools through improved Internet connections are Giga [23] and UNESCO's e-schools initiative. Giga, launched in (2019), is a joint initiative between International Telecommuni-cation Union (ITU) [24] and United Nations International Children's Emergency Fund (UNICEF) to connect every school to the Internet [1]. It exploits the power of meaningful connectivity to accelerate community members' access to educational resources and opportunities. UNESCO's e-schools initiative is another global effort highlighting the importance of connectivity for equitable and quality education. It aligns education sector investments with ICT policies to improve the achievement of learning outcomes and enhances employability for learners in various government sectors. UNESCO's e-schools initiative is based on a comprehensive Technology-Enabled Open Schools for All model, which includes a holistic framework encom-passing policy and resource enablers, technology, content, and human infrastructure; teaching, learning, and assessments for school connectivity programs. The model also advocates for leveraging available technology solutions to develop resilient school systems that provide continuous access to school education programs, despite phys-ical proximity. The e-school initiative also aims to support governments in creating a clear vision, developing school-wide programming and capacity-building strategies, and assessing results against specific targets to maintain an efficient and sustainable learning environment [1].

2.1.3   From ICT-Enhanced to Digital Transformation of Education

Digital transformation in education can be defined as digitizing educational processes and products to improve the teaching and learning experience for all [25]. It enhances accessibility to education and increases learners' engagement through customized and interactive learning. Intelligent classes, online learning, the Internet of Things, Augmented and Virtual Reality, Analytics, and Artificial Intelligence can be used to deliver improved and customized learning experiences. Digital transformation of education through ICT fundamentally requires a comprehensive multi-phased government plan based on the four pillars approach, which starts with mapping schools for connectivity, connecting schools, financing school connectivity, and empowering learners. The efforts of all stakeholders and the local community play a vital role in mapping and connecting schools. School connectivity equips learners with enhanced skills and knowledge, leading to more robust digital economics [1]. The policy of financing schools' connectivity should follow a holistic model that ensures suitable solutions based on specific connectivity requirements within the principles of affordability, operability, usage, financial viability, structure, and sustainability [15]. The focus on supply-side related challenges, demand-driven factors, and socio-cultural norms can contribute to the overall success of school connectivity plans as it is based on a comprehensive school connectivity approach. In a transformed educational system, the knowledge and expertise of all educators are devoted to transforming the traditional education system to achieve the SDGs [1]. This dynamic process involves implementing effective strategies, including mapping schools, digitally connecting schools, real-time monitoring of school connectivity, data sharing, financing school connectivity, and empowering learners [1].

According to a UNESCO report (2020), mapping schools for connectivity enhances transparency in monitoring the level of fulfillment in contractual obligations regarding connectivity and optimizes measurement of data around broadband Internet connectivity, schools, and the education system in general. The precise mapping of schools' location and connectivity can positively impact the efficiency of the delivered services, such as securing the equity and equality of educational opportunities and providing resources for the highest advantage, even in underserved areas. It also supports the obligatory measures needed to successfully execute the digitalization process at all levels. The school connectivity data and other datasets can be used to estimate the requirements and costs associated with expanding broadband connectivity to every class within the school, thus enabling efficient government financial and budget allocations. The same data can also be used during a national emergency for designing the national tender on compensations for educational loss and muster the funds to keep schools digitally connected. Furthermore, data sharing can support policymakers in estimating disparities in the current system and provide feedback that can be used to bridge the digital gaps [1].

The scale-up of real-time monitoring systems can highlight the challenges that could decelerate the entire connectivity process. It supports decision-makers in strategizing and supervising the whole school connectivity implementation processes [5]. Moreover, combining real-time data and Artificial Intelligence (AI)-based systems data can support the assessment of policies on the school's location, the number of students, type of broadband connectivity, and the impact of ICT on surrounding communities. Such assessments can assist the government in identifying the digital gaps, planning national policies, solving potential infrastructure problems, and mobilizing the required funds. For example, the United Nations Secretary General's High-Level Panel on Digital Cooperation [1] recommended an agenda for data-driven transformation by determining the techniques that can be used for extending the connectivity between schools. These techniques include reviewing the existing options for connectivity and analyzing the critical connection factors such as availability of intervention types, the required elements needed for the school's connectivity, regulatory framework, and suitable business model per the needs of students, administrators, parents, and instructors. Human and regulatory aspects must also be considered while executing the connectivity plan. The technology used for connectivity mostly depends on the need, usage, and context but must also provide affordable, fast, financially viable, and sustainable solutions [15].

All interventions must consider the needs of all stakeholders in the education system and ensure that accessibility requirements, such as equipment, proper training of staff, and child's online protection, are put into practice. Additionally, the cross-sectorial data must be shared with all ministries to provide the school with integrated solutions. The transparent, regulatory, and multi-sectorial connectivity models and interventions are designed to de-risk public–private investment. Therefore, financing school connectivity requires sustainable financial models based on ethical business standards [1]. Sustainable financial models reduce costs and generate higher revenues and new funding opportunities by focusing on associated risks and underserved areas. Investment in schools must consider the interests and needs of several stakeholders and adopt a transparent cost model that can reinforce reforms through a bottom-up approach. It is equally important to adopt a holistic approach by amalgamating private and public funding for digital connectivity and developing local entrepreneurship ecosystems that use private venture capital to empower the entire community.

2.2   A Case Study of Saudi Arabia Through the SABER-ICT Framework

The case of ICT implementation in education in Saudi Arabia is discussed using the *Systems Approach for Better Education Results* (SABER)-ICT framework [26]. SABER-ICT is a framework developed to cover eight ICT policy analysis domains: (1) vision and planning; (2) ICT infrastructure; (3) teachers; (4) skills and competencies; (5) learning resources; (6) EMIS; (7) monitoring and evaluation; and (8) equity, inclusion, and safety. Four evaluation phases for the policy development are identified for each sub-domain, including latent, emerging, established, and advanced. The framework supports benchmarking the baseline of policy development of a country, anticipating potential future policy directions, and gaining inspiration from other countries. In detail, the following sections will discuss the SABER-ICT approach linked to the Saudi case.

2.2.1 SABER-ICT

SABER-ICT is a World Bank initiative constructed on the systems approach to education analysis and reform to prepare comparative data on ICT in education policies [26]. SABER-ICT facilitates assessing the quality of education policies based on evidence-based global standards utilizing diagnostic tools aligned within each domain and across domains with detailed policy data, including governance, accountability mechanisms, information systems, financing rules, and school management. The framework is aimed at supporting policymakers make better-informed decisions regarding educational ICT. There are eight common themes and 20 critical areas of policy characteristics or sub-domains. For example, the domain vision and planning includes vision, policy linkages, funding, implementation authority, and private sector engagement. ICT infrastructure has power and infrastructure; the teacher domain provides for training, standards, support, and administrators; skills and competencies comprise digital competency and lifelong learning; learning resources include digital learning resources. The rest are education management information systems, monitoring, and evaluation, covering assessment, research, innovation, equity, inclusion, and safety, providing equity, digital ethics, and security. The framework is intended as a tool for analysis and guidance in benchmarking the current state of ICT/education policies to shape future directions and initiatives.

These thematic components are evolving analysis paths that can guide investments in educational ICTs and identify critical lessons during the application process. The author identified three sample use cases. The first use case is to analyze the current state of countries' ICT/education policy environment to propose potential next steps for improvement based on international best practices. The second is to project possible future policy directions. The third is to draw inspiration from ICT policies in other countries for national policy guidance. SABER-ICT aims to improve policy-related data, information, and knowledge related to using ICTs to enhance the quality of education [26] (Fig. 2.1).

2.2.2 ICT in Education: Saudi Arabia's Case

For ICT in education, Saudi Arabia implemented several projects and initiatives related to the connectivity, availability, affordability, and accessibility of government

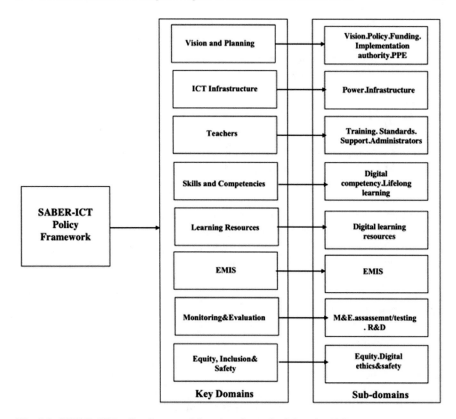

Fig. 2.1 SABER-ICT policy framework key domains and subdomains [26]

portals and services [27]. The National Transformation Program, one of the nation's Vision 2030 realization programs, identifies several objectives directly focused on the digital inclusion of all citizens, including vulnerable groups. These objectives include fostering values of equity and transparency, improving the quality of services provided to citizens, developing e-government, and strengthening communication channels with citizens and the business community [28]. The government identified two strategic objectives to support digital inclusion and equity: a digital service awareness campaign and digital service reach through intelligent government services [29]. Saudi Arabia improved digital connectivity by expanding the broadband infrastructure through a robust digital framework achieving 100% coverage for the mobile-cellular network population.

The International Telecommunication Union (ITU) provides a Digital Development Dashboard that reports the latest values for specific indicators from three ITU data sets, including data on access and use of ICTs by households and individuals, telecommunication and ICT infrastructure and access data, and price data. The dashboard provides an overview of the digital development state worldwide based on ITU data. The information is categorized into infrastructure and access, encompassing network coverage, mobile phone ownership, ICT access at home, and mobile and fixed broadband subscriptions. The second information category is Internet use which includes the percentage of the population using the Internet and broadband traffic. The third category consists of the enablers and barriers of digital development, including ICT prices and ICT skills.

Currently, 96% of the population is using the Internet, 97% of individuals own a mobile phone, 99% of households have Internet, 117% of active mobile broadband, and fixed subscriptions per 100 residents is 20% [30]. Saudi Arabia's Communications and Information Technology Commission (CITC) has also deployed free Wi-Fi hotspots in many public places, reaching 60,000 points nationwide and providing free access to government educational platforms [31]. Furthermore, the International Telecommunication Union [30] data show that 78% of Saudi Arabian individuals have at least basic ICT skills, 64% have standard ICT skills, and 14% have advanced ICT skills. The following sections discuss developments in educational ICT in Saudi Arabia following the SABER-ICT framework domains and subdomains.

2.2.2.1 Vision and Planning

The articulation and dissemination of a vision that considers the scope and scale of needed services are vital for guiding and leading ICT implementation efforts. Linking ICT policies to other policies across government agencies and ministries can also support the realization of the digital transformation vision. In planning, it is essential to consider regular and reliable funding mechanisms, authorize agencies and organizations to lead and oversee ICT implementation, and engage with the private sector for funding. These actions can prompt access to industry experts and promote stakeholder coordination [26]. As identified by Kozma [32], strategic policies supporting ICT integration in education are based on one or a combination of the following: (1) to stimulate economic growth, (2) to enhance social development, (3) to accelerate education reform, and (4) and improve education management. The national ICT vision and goal in Saudi Arabia were foregrounded with the launch of the Watani project (2000), the Saudi schools' net project implemented by the Ministry of Education. The project aimed at leveraging ICT in education through professional development for teachers and support for learners to improve their digital skills. Further, it creates a digital learning culture responsive to different learning needs and prepares learners to be active participants in the knowledge economy (MoE, Kingdom of Saudi Arabia 2000). To achieve these goals, Saudi Arabia invested in educational ICT to prepare learners for the future workforce and to be better participants in the national economic growth [33, 34]. In (2005), the National e-Learning Center was established, followed by the Tatweer Company for Education Services (T4edu) in (2012), with a critical goal to reform educational outcomes through increased integration of technology in public schools. In (2016), the Tatweer Educational Technologies Company (TETCO) was established to expand the use of ICT to enhance the quality of education and teachers' digital skills, develop digital curricula, enhance schools' activities utilizing ICT, and improve the whole school environment. These projects were followed by launching the Education Information Network (EiN) Portal [35], which hosts course materials, recorded lessons, enrichment activities, question banks, and professional development resources and content for teachers to support the MOE goals.

The National Transformation Plan (NTP; 2015–2020), one of Saudi's Vision 2030 realization programs launched by the Ministry of Economy and Planning, lists 24 objectives that prioritize actions for developing ICT skills and technology integration in education. These objectives highlight the need to upskill teachers and learners in ICT [36]. This national program aims to build a future-oriented digital government, establish a thriving digital economy based on the Fourth Industrial Revolution trends, and create a vibrant digital society. The NTP's general objectives include 'improving teachers' recruitment, training, and development, improving the learning environment to stimulate creativity and innovation, and improving curricula and teaching methods [37]. One of the MoE's critical objectives of the NTP is to sustain the 'shift to digital education in supporting both teachers and students and increase the private sector's participation in education and training ([32], p. 38). Furthermore, the focus of the 10th Saudi National Development Plan (2015–2019) was to support a knowledge-based economy by promoting 'the main determinants of productivity and economic growth through focusing on information technology and education to achieve a distinct economy' ([36], p. 38). In addition, the Saudi National Transformation Program (2020) aims to equip young Saudis with essential digital skills to live and work in a global economy [37].

The immediate implementation of distance education on a large scale during the COVID-19 pandemic can be attributed to Saudi Arabia's proactive investments in ICT, significant aggregated experience in educational technologies, and well-coordinated governance efforts across multiple entities [38, 39]. Leadership efforts, quick decision-making, and expertise built from prior investments in digital education were the critical success factors that supported the rapid transition to online learning [40]. Since the COVID-19 Pandemic, the demands on technology have increased exponentially, especially for a capable centralized infrastructure that can manage high-volume use, ensure security related to student data and privacy, and improve internet access, reliability, coverage, and network [41]. These recommendations are further confirmed in the OECD report [42], *How the COVID-19 Pandemic is Changing Education: A Perspective from Saudi Arabia.* The report focuses on Saudi Arabia compared to international data to explore the country's preparedness to deal with the Pandemic consequences and ensure educational continuity. The report shows that many students may need more resources for digital education.

Educational leadership readiness for ICT integration appears to be on the same pace as online learning development and delivery in a pre-COVID environment. Critical factors related to administration, such as appropriate governance, strategic national plan alignment with other strategic plans within the school or district, and enacted and continuously reviewed policies, were in place and undergoing improvement. Adequate resources were projected to sustain and scale online learning and prepare sufficient qualified professionals and administrative and support staff. Nevertheless, alignment and implementation of online education at the strategic planning level, the inclusion of policies and processes, and ample resources and human resources to support a fully online environment are still needed [43]. The World Bank report [43], Saudi Arabia's Digital and Distance Education: Experiences from the COVID-19 Pandemic and Opportunities for Educational improvement, explores Saudi Arabia's readiness for digital and distance education following the COVID-19 pandemic identifying the strengths and opportunities for future improvements. Table 2.1 represents the vision and planning domain and subdomains ([26], p. 8) with the current state of the respective subdomain based on the literature.

Table 2.1 The vision and planning domain and subdomains' current stage based on literature [26]

1-Vision and planning		
Subdomain	Stage	Literature
1.1 Vision and overall goals	• ••• "Explicit policy guidance related to ICT/education topics; ICT in education policy is fully operationalised and seeks to transform learning environments, teaching practices, and administrative processes with the aid of ICTs"	Alghamdi and Holland [27] Online Learning Consortium OLC [34]
1.2 Linkages between ICT in education policy and other sectoral policies	• ••∘ "Many linkages between ICT policy and other education policies"	MoEP, Kingdom of Saudi Arabia [36] World Bank [43]
1.3 Public funding and expenditure for ICT in education	• ••• "Regular public expenditure on ICT in education on infrastructure and many no infrastructure items"	Alghamdi and Holland [27] Albugami and Ahmed [33] Online Learning Consortium OLC [34]
1.4 Institutional arrangements	• ••• "Dedicated, professionally staffed focal unit/agency charged with implementing policies on ICT in education actively coordinating with other organizations"	[35] Online Learning Consortium OLC [34]
1.5 Public Private Partnerships (PPP) on ICT in education	• ••• "Explicit commitment to integrating, coordinating and monitoring PPP initiatives related to ICT/education"	Ministry of Education [35] Alshehri et al. [38] Barfoot [39] Online Learning Consortium OLC [34]

2.2.2.2   ICT Infrastructure

Regarding infrastructure, the SABER-ICT framework highlighted the importance of ensuring adequate infrastructure through reliable and affordable access to power, particularly for rural and underserved areas. The framework also focuses on the importance of providing sufficient equipment, networking infrastructure, and technical support to meet the demands for access to ICT devices and reliable Internet connectivity. In Saudi Arabia, the Schools Connectivity Project (2013) in KSA is a part of a large nationwide connectivity plan to connect schools across the country. The project goal was to connect over 3000 small schools with no Internet Service using Satellite and to enhance and upgrade the available connectivity in more than 19,000 schools with wired and wireless connectivity to Direct Internet Access (DIA). This strategic project aimed at linking schools to MoE centralized services such as Faris (School Teachers and Administrators Management System) and Noor (Students Identification System) [27]. The long-term goal is to have a reliable, expandable infrastructure that meets the respective connectivity needs of each school. The project execution and implementation were gradual through multiple phases to cover different regions and meet budget allocation plans to ensure schools' readiness and capacity building during the implementation process. The generated data from piloted projects supported subsequent phases of project implementation. This project highlights the importance of an iterative planning approach that offers constant support for clients through quality service mechanisms and monitoring processes. It also reiterates the need for partnerships with regulatory firms and the private sectors, such as the communication industry.

National plans and efforts are working on increasing access to centralized online learning programs among the ICT infrastructure by identifying schools that have not yet implemented online learning and centralized Learning Management Systems LMS and Single Sign-On (SSO) [34, 43]. The OLC [34] reports the state of Saudi Arabia's online K-12 education and envisions the post-COVID environment to explore opportunities for improvement and highlight the evaluation framework for quality K-12 online learning by surveying 385,957 members of the Saudi education community.

The MoE identifies schools that have not implemented online learning to provide centralized LMS with SSO. For operability, a feedback loop was developed to ensure parents and students' needs were met and that centralized and comprehensible policy resources were made accessible for them. For security, clear data security plans and policies around using third-party applications were developed and shared with stakeholders. Regarding Information Technology Service Management (ITSM) compliance, the current processes, procedures, and resources were evaluated, and new resources to address challenges were developed. The MoE is currently in the process of demining IT access challenges for stakeholders. For reliability, the MoE developed methods for monitoring and addressing the reliability of the centralized infrastructure and learning technology systems. In addition, stakeholders were provided with multiple ways to respond to needs and follow up on the market for new technologies [34].

   Continuous coordination between MoE and the Ministry of Communications and Information Technology was undertaken to improve Internet quality in all regions and reinforce the educational system's readiness for connectivity. MoE also collaborated with other ministries and public agencies, such as the Ministry of Finance, the National Cybersecurity Authority, the National eLearning Centre, and telecommunication companies, to provide adequate technology resources. These partnering entities worked together to enhance the ability of the infrastructure to support high-volume usage and provide centralized access to online learning. The available network was also adequately powered to handle the demands of online teaching and learning, either at schools or at home. Table 2.2 represents the ICT Infrastructure domain and subdomains ([26], p. 9) with the current state of the respective subdomain based on the literature.

Table 2.2 The ICT Infrastructure domain and subdomains' current stage based on literature [26]

2- ICT Infrastructure		
Subdomain	Stage	Literature
2.1 Electricity supply	• ●●● "School electrification is not an issue"	Ministry of Education [35]
2.2 ICT equipment and related networking infrastructure	• ●●○ "All schools have reliable broadband Internet access and computing facilities Some computers in classrooms Widespread use of digital projection equipment Some teachers are provided with computers or related devices. Computing facilities are being introduced at scale in libraries, science labs, and/or other venues inside schools Some e-Waste policies are in place"	World Bank [43] Alghamdi and Holland [27] Ministry of Education [35]
2.3 Technical support and maintenance	• ●●○ "Generally sufficient, systematic technical support mechanisms in place Maintenance costs are budgeted regularly"	Online Learning Consortium OLC [34] Ministry of Education [35]

2.2.2.3   Teachers

For the teacher's domain, the SABER-ICT framework highlighted the importance of providing teachers ICT-related training (technical and pedagogical) and the significant impact of increased investment in professional development [26]. It also emphasized the importance of identifying teacher ICT competency standards, providing support for teachers in their use of ICT, and the value of raising awareness among teachers and school administrators on the importance of ICT in education [26]. The MoE worked on creating a centralized support space for teachers through the unified educational portal, Madrasati [43]. School administrators shared and promoted the resources from this portal in the school environment. MoE developed continuous professional and technology-related training and support for teachers to develop their competencies in designing digital learning experiences and teachings in technology-rich classrooms, such as the Back-to-School website for professional development in ICT knowledge and skills. Extended capability-building opportunities were provided to teachers [34]. In addition, MoE developed and shared digital teaching orientation modules for schools and designed virtual peer support and mentoring programs for collegial learning groups. Through a partnership with the National Center for Professional and Educational Development [34, 44], MoE provided training to 428,318 teachers to support its digital transformation in education plans. Teachers exchanged best practices in teaching with technology in MoE online teachers' communities of practice groups. Teachers received training and support through several other initiatives such as intensive training of trainers' programs, professional development webinars on different digital teaching skills, skills development workshops, and online conferences with education supervisors on how to teach using digital tools and resources [34, 43]. Table 2.3 represents the teacher's domain and subdomains ([26], p. 10) with the current state of the respective subdomain based on the literature.

2.2.2.4   Skills and Competencies

For skills and competencies, the framework focuses on identifying ICT literacy competency standards and promotes the importance of offering related training, support, assessment, and certification associated with the backing for ICT-enabled lifelong learning opportunities [26]. Regarding digital competency, one of the initiatives that MoE developed and implemented was comprehensive online learning orientations and academic support resources for students [34, 35]. This initiative allowed students to engage with other members with digital citizenship knowledge, skills, and values. School administrators shared and promoted the resources from this portal in the school environment. MoE developed continuous professional and technology-related training and support for teachers to develop their skills in designing digital learning experiences and teachings in technology-rich classrooms, such as the Back-to-School website for professional development in ICT knowledge and skills. Extended capability-building opportunities were provided to teachers

Table 2.3 Teacher's domain and subdomains current stage based on literature [26]

3-Teachers

Subdomain	Stage	Literature
3.1 Teacher training and professional development (including pre-service and in-service) on ICT—related topics	• ••• "ICT use and related pedagogical training and support is integral to regular, on-going programs aimed at providing targeted pre-service and on-going professional development opportunities for teachers"	World Bank [43] Online Learning Consortium OLC [34]
3.2 ICT-related teacher competency standards	• ••• "ICT-related competency standards for teachers and related certifications are in place and integrated into general competency standards for teachers"	UNESCO [44] Online Learning Consortium OLC [34]
3.3 Teacher networks/resource centers for teachers	• ••• "Online and/or mobile teacher support networks in widespread use ICT-related resource centers are funded and play important positive roles in supporting teachers"	Online Learning Consortium OLC [34]
3.4 School leadership training, professional development, and competency standards	• ••• "ICT-related training for school leaders is readily available, and related competency standards are in place"	Online Learning Consortium OLC [34]

[34]. In addition, MoE developed and shared digital teaching orientation modules for schools and designed virtual peer support and mentoring programs for collegial learning groups.

Furthermore, MoE launched two competitions for students in coding and in digital content authoring named (Madrasati Coding), where infographics, podcasts, videos, and motion graphics were used to create and curate online content. These competitions were designed to disseminate a culture of programming and innovation among students, reinforce students' and teachers' digital skills, increase parents' participation in their children's education, and encourage innovation and programming-based practices. The number of participants in these competitions exceeded 4,734,183 students (Saudi Ministry of Education 2022). Table 2.4 represents skills and competencies' domain and subdomains ([26], p. 11) with the current state of the respective subdomain based on the literature.

Table 2.4   Skills and competencies' domain and subdomains current stage based on literature [26]

4. Skills and competencies

Subdomain	Stage	Literature
4.1 ICT literacy/digital competency	• ••• "Digital competency is viewed as an essential 21st-century skills ICT-related competency frameworks defined beyond just technical skills"	Ministry of Education [35] Online Learning Consortium OLC [34]
4.2 Non-formal education/lifelong learning/vocational education	• ••• "There is an robust integrated vision for lifelong learning enabled by ICT"	World Bank [43] Ministry of Education [35]

2.2.2.5   Learning Resources

The SABER-ICT framework highlights the importance of supporting the development, sharing, and implementation of digital learning resources. MoE launched Madrasati, a nationally developed learning management system, in (2020). While a previous iteration of the platform has been in use since 2014, the COVID-19 pandemic accelerated the platform's implementation on a national scale. Madrasati is a unified e-learning platform with links to tools and services for all users, including students, teachers, school principals, and parents. Madrasati is developed as a virtual school model to provide an online learning environment that simulates the in-person school experience and model with various instructional processes and activities. These include Microsoft (MS) Teams, Office 365, iEN National Gate Portal, and interactive tools [43]. The Madrasati platform hosted over 154 million virtual classes and over 16 million educational resources, including augmented reality lessons, videos, games, interactive three-dimensional objects, interactive educational stories, and books. This platform supports synchronous and asynchronous educational approaches, with more than 700 million test tutorials and simulations and 41 million instructional activities created by subject matter teachers. The platform supports teachers in implementing quality assessment through free access to question banks that include over 93,000 valid and reliable questions for most subjects (Saudi Ministry of Education 2022).

In addition, MoE launched iEN National Gate, a free portal for a repository of educational materials, which was developed in 2015 in partnership with T4edu. This portal provides an educational experience and content management system applying international quality standards and best practices. The portal provides multiple educational resources such as digital books, e-tests, and self-assessment tests, recorded lessons, lesson plans (over 450,000 lesson plans created with teachers' support), instructional design guidelines, and a variety of educational games, videos, and content resources based on virtual and augmented reality and three-dimensional

(3D) technology. This portal is integrated into Madrasati platform to support seamless access to digital content through Madrasati. The iEN portal system and content are regularly updated. MoE also launched an iEN 23 T.V. Channel that broadcasts free lessons online for all K-12 students and teachers. More than 230 million views were reached within a year, 24 million hours watched, 186,000 h of satellite broadcasting, and over 25,000 h of filming had been accomplished through this channel [43]. Another learning resource is Virtual Kindergarten (2019), an educational system for children between three and six years old. It simulates the reality of kindergarten through educational videos, stories, and games. During the first semester of 2020–21, the Virtual Kindergarten application had over 3.5 million views. By the spring of 2020, more than 300,000 children and 283,000 parents had registered on this application [45].

The Back-to-School website offers more than 300 educational and instructional materials [43]. The content provides instructional guides, videos, and infographics that can be accessed through the Madrasati platform. Users are instructed in how to use the Madrasati platform effectively, apply quality e-learning successfully, and apply online teaching strategies. Back-to-School is designed for all platform users, including kindergarteners and students with disabilities. Various guides and infographics have been developed and made available on the Back-to-School Platform (backtoschool.sa) for different users (students, parents, school principals, teachers, educational supervisors, kindergarteners, and people with disabilities). Table 2.5 represents learning resources' domain and subdomains ([26], p. 12) with the current state of the respective subdomain based on the literature.

2.2.2.6 EMIS (Education Management Information Systems)

The SABER-ICT framework emphasized the importance of collecting, processing, analyzing, and disseminating education-related data to relevant stakeholders. Educational ICT policies require systematic and holistic data views to include education management information systems within the broader ICT/education policies [26]. Data-based learning analytics equip educators with the power to design more effective student-centered teaching plans tailored to accommodate student strengths and challenges [46–48]. Majumdar et al. [49] stated that learning analytics facilitated generating prescriptive reports for students based on the engagement score and supported generating sophisticated prescriptive algorithm-based and data-driven analytics.

In this regard, MoE developed Noor, an educational management system that provides a centralized database to ensure the rapid exchange of information and resources for more than 10 million users, including 6 million students and half a million teachers working in over 40,000 public and private schools in Saudi Arabia [27]. Noor provides online access to performance and historical records. It serves planning, control, and decision support systems through data mining. Noor efficiently facilitates a path for the Electronic Document Delivery System (EDSS) buildup and provides dynamic 'power query' reporting mechanisms for customized (EDSS)

Table 2.5 Learning resources domain and subdomains current stage based on literature [26]

5-Learning resources		
Subdomain	Stage	Literature
5.1 Digital content/digital learning resources (DLR) and curriculum	• ••• "Teachers and students access digital content/DLR resources linked to specific curricular and learning objectives, anytime anywhere, using a range of devices and platforms Teachers and students regularly use, develop- reuse and re-mix develop digital teaching and learning resources Advanced' digital learning resources (e.g., robotics, simulations, games) are used in teaching Intellectual property issues are well considered Users are involved in evaluating and assessing the quality of digital learning resources"	Saudi Ministry of Education (2022) World Bank [43]

statistical outputs and dynamic KPIs. It seamlessly integrates Learning Management System (LMS), Student Information System (SIS), Assessment Management System (AMS), and Communication and Collaboration Tools (CCT) in one accessible Cloud hosted solution. Since Noor is based on learning standards, it is preloaded with Common Core Standards-CCSS and other international standards [50]. Table 2.6 represents the education management information systems (EMIS) domain and subdomains ([26], p. 13) with the current state of the respective subdomain based on the literature.

2.2.2.7 Monitoring and Evaluation, Assessment, Research, and Innovation

The SABER-ICT framework emphasized the importance of monitoring ICT use in education and evaluating its impact on learning and teaching. It recommends using ICT for assessment activities and dedicating support to innovative uses of ICTs in education [26]. MoE developed processes and policies on course and program evaluation, including reviews and updates to gather feedback from students and parents. In addition, frequent feedback from stakeholders was collected, and common issues were resolved to keep the Madrasati platform applicable and receptive. Students and parents were informed regarding the impact of their feedback on the development of digital education. Software updates were also regularly implemented. MoE

Table 2.6 Education management information systems (EMIS) domain and subdomains current stage based on literature [26]

6-Education management information systems (EMIS)		
Subdomain	Stage	Literature
6.1 ICT use in the management of the education system (EMIS)	• ••• "National EMIS uses ICT to collect, process, and store information produced by various levels of the education system and to disseminate data to various levels of the education systems as a critical decision tool, accessible via multiple channels (e.g., Internet, mobile devices), with some access to EMIS data available to the general public and with robust security and data privacy safeguards in place"	Alghamdi and Holland [27] ITG [50]

created a communication plan regarding evaluation opportunities and purposes to ensure teachers and staff's satisfaction. In addition, MoE formed senior and steering committees with cascading levels for educational administration on a regional level, administrative level, district level, and school level. The committees ensure that ICT policies and frameworks are effectively delivered in all 47 regional educational directorates. They also evaluate and monitor the implementation of ICT initiatives in schools in each region [35].

In addition, MoE provides support through several channels, including the Tawasul support service (24-h hotline between teachers, students, parents, and the Ministry), electronic tickets, and support email [35]. Support includes a call center, online live chat, support staff linked to district offices, and online guidance information for schools on the log-in queries. The 23 iEN channels adopt a continuous evaluation process to ensure the quality of recorded lessons before going live on iEN YouTube channels and be uploaded to the Madrasati platform as part of either synchronous or asynchronous instructional content.

MoE, in partnership with the National Center for Educational Professional Development (NCEPD), has conducted several comprehensive studies regarding research and innovation endeavors. The Online Learning Consortium [51] proposed 71 development initiatives as recommendations for evaluation with the participation of the International Society for Technology in Education (ISTE), Quality Matters (QM), the UNESCO Institute of Information Technologies in Education (IITE), the National Research Center for Distance Education and Technological Advancements (DETA) in the USA. The recommendations included eight overarching dimensions and several subdimensions, including leadership, digital curriculum design, e-teaching and learning, learning evaluation, technology, support training, continuous development, and assessment, which were studied to understand the national-level progress

Table 2.7 Monitoring and evaluation, assessment, research and innovation domain and subdomains current stage based on literature [26]

7. Monitoring and evaluation, assessment, research, and innovation		
Subdomain	Stage	Literature
7.1 Monitoring ICT use and evaluating its impact	• ••• "Robust M&E system is in place to measure the use of ICT across a variety of areas and related impacts, including learning outcomes Policy choices and decisions related to ICT informed by rich evidence base M&E function or activities related to ICT/education largely independent of project implementers"	Ministry of Education [35] Online Learning Consortium OLC [34]
7. 2 Innovation, research and development	• ••• "Special budget or fund for R&D activities Centers of excellence for R&D of ICT use and services R&D and innovation are central to planning—it is part of the system"	Online Learning Consortium OLC [34]

of educational ICT. The MoE also plans to develop processes for monitoring the reliability of ICT infrastructure, learning technology systems, and courses (OLC 2022). Table 2.7 represents the monitoring and evaluation, assessment, research, and innovation domain and subdomains ([26], p. 14) with the current state of the respective subdomain based on the literature.

2.2.2.8 Equity, Inclusion, and Safety

The SABER-ICT framework prioritizes pro-equity educational ICT provisions and approaches for specific marginalized groups. The framework also highlights articulating and supporting ethical practices that include provisions that ensure data confidentiality, protection, and security [26]. In terms of equity, MoE provided students with online support to ensure equal and effective support opportunities. The accessibility standards for digital learning environments were reviewed to address potential gaps in accessibility and to ensure compliance standards. Further, MoE made efforts to enhance equity and inclusion for all students by collaborating with charity institutions such as Takaful to provide devices and free Internet packages for underserved and low-income students. In addition, telecommunication companies provided trucks to strengthen signals in areas with low connectivity. Further, MoE collaborated with the private sector to provide free online education applications, services, and discounted internet packages [35].

According to the Security Scorecard report (2018), the education sector is ranked last among 17 cybersecurity-prepared industries and performed poorly in patching cadence and application and network security [52]. Therefore, for security measures, MoE enhanced its existing firewalls to prevent unauthorized access to data via breaches or leaks. Data protection also includes user authentication that supervises users' access. Students' confidentiality was protected by implementing educational platform data governance, privacy, and usage policies. To ensure accessibility, the MoE applied accessibility policies and standards (e.g., The Web Content Accessibility Guidelines: WCAG 2.0) to its different learning resources. For example, sign language for students with hearing impairment was implemented on iEn YouTube channels.

Similarly, thousands of recorded lessons provide sign language options on the platform. Three open educational channels (on iEN and YouTube) provide personalized curricula for students with special needs. Further, multiple accessibility options and supportive digital tools were offered to children with special needs through the Madrasati platform and Microsoft (MS) Teams [35]. Table 2.8 represents the equity, inclusion, and safety domain and subdomains ([26], p. 15) with the current state of the respective subdomain based on the literature.

This section analyzes the current state of Saudi Arabia's ICT/education policy environment to propose potential next steps for improvement based on international best practices as one of the use cases of the SABER-ICT framework [26]. The second

Table 2.8 Equity, inclusion, and safety domain and subdomains current stage based on literature [26]

8. Equity, inclusion, and safety		
Subdomain	Stage	Literature
8.1 "Pro-equity" provisions and approaches	• ••○ "Most of the following "proequity" provisions or approaches related to ICT use in education are addressed: gender equality; rural/urban divides; low-income communities; special needs students; indigenous groups; gifted students older learners"	Ministry of Education [35] Online Learning Consortium OLC [34]
8.2 Digital ethics, safety and citizenship	• ••• "Legislation covers all aspects of digital safety, issues are digital ethics and citizenship are integrated into considerations of ICT use"	Ministry of Education [35] Online Learning Consortium OLC [34]

sample use case is to project possible future policy directions. The following section provides recommendations and potential future improvement directions.

2.3   Recommendations

Digital transformation of the learning and teaching processes can create a dynamic and engaging educational experience for schools. Governments must adopt a holistic digital transformation approach to innovate and develop smart classrooms by disseminating a transformation culture, maintaining continuous and effective leadership, leveraging cyber security and ICT infrastructure, and exploiting the power of learning analytics. Digital transformation aims at responding to the ever-growing demands of students, teachers, and schools by creating an ecosystem that combines technology, services, and security to bridge the digital gap and create collaborative, interactive, and personalized learning experiences.

2.3.1   Transformation Culture

Significant changes in the educational system require an ambitious and shared vision that focuses on improving and innovating school practices, classroom instructions, and technology-related decisions. To achieve effective digital transformation, school leaders must embrace and disseminate a culture of change that ensures digital equity and accessibility to learning resources and digital content. Big data solutions can offer a customized learning experience for students, enabling diagnosis of learning challenges and offering a personalized learning experience. Effective digital learning platforms and LMS foster knowledge exchange exponentially beyond classroom constraints and allow for globalized dissemination of knowledge and internationalized educational community conversations. Another essential factor of transformation is teacher empowerment. Professional development programs need to equip teachers to be active change agents to sustain meaningful change and transformation. Digital transformation and disruption are opportunities for school ICT teams to operate and deliver scalable and secure solutions across diverse networks and devices. Therefore, well-planned and relevant professional development opportunities combine the latest digital practices with job-embedded coaching programs.

2.3.2   Policy and Leadership

School leaders committed to the school transformation process envision creating technology-integrated and 21st-century learning environments. They have clear and defined goals, implementation plans, and action steps. Transformative educational leaders should encourage innovative blended learning models and flipped classroom policies and models, where the traditional order of instruction is reversed, and recorded video presentations are part of the instruction process. Research, development, and innovation endeavors must be maintained to help school leaders use technology effectively since school leaders play a crucial role in sustaining and improving digital transformation initiatives. Leadership and policy efforts should also consider making the right technology decisions regarding the most viable options for technology devices and connectivity choices. In addition, providing user experiences that are intuitive and consistent with learning outcomes is another crucial goal for ICT departments. Customer Relationship Management (CRM) systems are another vital asset for schools since they allow schools to manage the entire lifecycle of a potential customer, including communication and relationship with school members. Thus, they provide the best customer experience by adapting customer relationship strategy to their needs and preferences.

Further, the one-to-one computer program approach seeks to provide laptop computers and Internet access to students at home and school, providing laptops for all students, creating mobile computer labs with one lab for each student and supporting Bring Your Own Device BYOD initiatives. It is also essential to plan budget and funding allocation for infrastructure. Expenditures must be aligned with the established educational goals and technology access students have outside school.

Digital transformation policy must be aligned and linked to educational goals by focusing on desired learning outcomes and academic and institutional purposes. District leaders must develop streamlined technology policies for administrators, staff, faculty, and parents. Quality assurance policies play an integral role in comprehensive educational transformations and must be in place to monitor and evaluate practice and leadership efforts. Partnerships and cooperation between public and private education sectors are vital for enhancing enrollments and student outcomes and managing costs effectively. Data analytics support formulating effective educational policies and academic decisions. Leaders must deploy an educational transformation framework with a strategic plan that can leverage organizational capacity and sustainability with crucial performance and success factors supported by the Acceptable Use Policy (AUP).

2.3.3 Security and Cybersecurity Readiness

The vast amount of collected data on students and learning processes highlights that cybersecurity and network security needs to be prioritized by both school faculty and decision-makers. Data security plans and procedures must be a dynamic and ever-evolving process. Educational leaders need to develop a comprehensive security strategy that includes the most effective and best practices for cyber security. These practices should establish safe learning environments with minimum cyber risks, multiple levels of network security, and a zero-trust ecosystem. School leaders must create a culture of security and data privacy and support policies consistent with administrative regulations and day-to-day practices to promote responsible use of technology. In addition, all educational institutions must adopt a holistic approach to student data protection. Therefore, a robust cybersecurity plan can be implemented to cover processes related to monitoring and managing networks, maintaining and upgrading equipment, estimating network capacity, and leveraging protection with firewalls and anti-virus software. A cyber security plan should implement network redundancy and backup recovery plans and align with regulatory requirements for cybersecurity performance to improve the education industry.

2.3.4 Infrastructure

Developing technological infrastructure is a fundamental step toward digital transformation in educational organizations, and academic leaders must create value-added strategic initiatives to help school districts achieve their goals. Indeed, there is a significant connection between access to broadband and student performance. Solid, resilient, and cost-effective infrastructure solutions must be implemented to ensure a smooth rollout of school educational technology initiatives. Such initiatives will also improve performance, increase schools' flexibility, and help them better respond to various educational needs with minimal downtime, disruption, or performance gaps.

The next generation of hyper-converged infrastructure solutions must adapt to the growing and changing educational needs, environments, and data requirements. Robust and efficient infrastructure could provide seamless connectivity for many devices and applications. Additionally, investing in Cloud computing enables schools and districts to adapt immediately in times of crisis and maintain education continuity at scale. Schools must ensure students' cybersafety by investing in firewalls, content filters, and virtual meeting safety measures to block pervasive phishing and malware threats regarding sensitive school data and other online safety incidents. Cybersecurity and Smart Education Networks by Design (SEND) initiatives at the Consortium for School Network (CoSN) (2020) point to three main areas of concern for schools. School networks, their critical services, and unmanaged software can be vulnerable to potential attacks, primarily if they have also to administer devices running on home networks. Therefore, integrated cloud security solutions are required for supporting IT and security teams. Such integrated systems will have better traffic visibility and can effectively control applications and data flow across networks and multi-cloud environments. The essential requirements of the school cloud security solution include data loss prevention, malware, and threat protection, user monitoring, and content scanning.

Data centers' infrastructure is another critical element of schools' digital infrastructure. Data centers handle taxing workloads during virtual learning, new technology integrations, and data analytics. Further, hyper-converged infrastructure is the fastest growing IT data management solution for separate centralized systems such as data computing, storage, and networking. This infrastructure can save time and resources since it eliminates silos, simplifies management, and enables scalability without sacrificing performance or availability.

2.3.5 Learning Analytics

Learning analytics has the potential to transform education by supporting the personalization of education and enhancing the teaching and learning process. Data-based learning analytics equip educators with the power to design more effective student-centered teaching plans tailored to accommodate student strengths and challenges. Learning analytics can support educators in identifying specific learning patterns, such as monitoring students who need extra support, registering students' engagement, measuring and comparing students' performance, diagnosing learning difficulties, and redesigning courses and learning activities accordingly. It enables monitoring of learning progress and facilitates necessary interventions for students who need additional assistance. Several important recommendations are to be considered by school leaders while implementing learning analytics. For example, schools can adopt a methodological process to evaluate needs and context, such as using Supporting Higher Education to Integrate Learning Analytics (SHEILA) [53] as a framework that can guide the adoption of learning analytics. The framework offers an instrument for engaging stakeholders in documenting actions, policy directions, and potential challenges.

Nevertheless, ethical issues related to students' age, type of data analysis, and source of data are all critical concerns that need to be considered. Finally, learning analytic techniques must focus on the learning process and outcomes. Fourth, combining deep learning with explainable artificial intelligence techniques is essential for analyzing educational data.

2.4 Conclusion

To conclude, the digital transformation of education through ICT fundamentally requires a comprehensive multi-phased government plan based on a systematic and multifaceted approach. The efforts of all stakeholders and the local community play a vital role in schools' digital transformation and disruption. School leaders committed to the school transformation must formulate a clear vision for creating technology-integrated 21st-century learning environments. This vision needs to be supported with clear goals, implementation plans, and action steps to prepare globally competitive citizens and create more robust digital economies. Further, school leaders need to consider critical elements supporting digital transformation plans such as funding efficiency, private sector engagement, ICT infrastructure robustness, teacher training effectiveness, equity and security standards, digital competencies, and lifelong learning opportunities and resources sustained by research and innovation. Improved education quality in schools and enhanced level of digital knowledge and skills, including skills related to e-government and other digital service platforms, significantly rely on ICT implementation in schools.

The third sample use case of the SABER-ICT framework is to draw inspiration from ICT policies in other countries for national policy guidance to improve the availability of policy-related data, information, and knowledge related to using ICTs to enhance the quality of education. The framework is intended as a tool for analysis and guidance to benchmark the current state of ICT/education policies in countries and shape future directions and initiatives. The framework is developed to help policymakers make informed decisions regarding educational ICT. The thematic components are evolving analysis paths that can guide investments in educational ICTs and identify critical lessons learned during the application process. Thus, international multilateral analysis of ICT-related educational policies enhances exchanging lessons learned globally and encourages the dissemination of best international practices.

References

1. UNESCO. (2020). The digital transformation of education: connecting schools, empowering learners. Available at: https://unesdoc.unesco.org/ark:/48223/pf0000374309. Accessed 17 June 2022.
2. ITU/UNESCO Broadband Commission for Sustainable Development. (2019). The State of Broadband: Broadband as a Foundation for Sustainable Development. [online]. Available at: https://www.itu.int/dms_pub/itu-s/opb/pol/S-POL-BROADBAND.20-2019-PDF-E.pdf. Accessed 17 June 2022.
3. World Economic Forum. (2019). Schools must look to the future when connecting students to the Internet. [online]. Available at: https://www.weforum.org/agenda/2019/02/schools-must-look-to-the-future-when-connecting-students-to-the-internet/. Accessed 17 June 2022.
4. Kingdom of Saudi Arabia. Saudi Vision 2030. (2016). https://vision2030.gov.sa/en. Accessed 16 June 2022.
5. Khalil, N., Ali, U., Cemil, K., & Friedrich, S. (2022). e-Government, education quality, internet access in schools, and tax evasion. *Cogent Economics & Finance, 10*(1), 2044587. https://doi.org/10.1080/23322039.2022.2044587
6. Baum, C., & Di Maio. A. (2000). Gartner's four phases of e-government model. http://www.gartner.com/DisplayDocument?id=317292. Accessed 10 August 2022.
7. Adam, I. O. (2020). Examining e-Government development effects on corruption in Africa: The mediating effects of ICT development and institutional quality. *Technology Society, 101245*(61), 1–10.
8. Zhang, Y., Liu, X., & Vedlitz, A. (2020). Issue-specific knowledge and willingness to coproduce: The case of public security services. *Public Management Review, 22*(10), 1464–1488.
9. World Bank. (2018). World Development Report. Learning to realize education's promise. Available at: https://www.worldbank.org/en/publication/wdr2018. Accessed 17 August 2022.
10. Marthandan, G., & Tang, C. (2010). Information technology evaluation: Issues and challenges. *Journal of System and Information Technology, 12*, 37–55.
11. Twizeyimana, J. D., & Andersson, A. (2019). The public value of e-Government—A literature review. *Government Information Quarterly, 36*(2), 167–178, ISSN 0740-624X. https://doi.org/10.1016/j.giq.2019.01.001. Accessed 20 June 2022.

12. Malodia, S., Dhir, A., Mishra, M., & Bhatti, Z. (2021). Future of e-Government: An integrated conceptual framework. *Technological Forecasting, and Social Change, 173*, 121102, ISSN 0040-1625. https://doi.org/10.1016/j.techfore.2021.121102. Accessed 18 May 2022.
13. Meiyanti, R., Utomo, R., Sensuse, D., & Wahyuni, R. (2018). e-Government challenges in developing countries: A literature review. In *2018 6th International Conference on Cyber and IT Service Management (CITSM)*, pp. 1–6. https://doi.org/10.1109/CITSM.2018.8674245. Accessed 4–9 August 2018.
14. Wu, N., & Liu, Z. (2021). Higher education development, technological innovation, and industrial structure upgrade. *Technological Forecasting and Social Change, 162*, 120400.
15. Chen, Y., Mittal, V., & Sridhar, S. (2021). Investigating the academic performance and disciplinary consequences of school district internet connection spending. *Journal of Marketing Research, 58*(1), 141–162.
16. Shakina, E., Parshakov, P., & Alsufiev, A. (2021). Rethinking the corporate digital divide: The complementarity of technologies and the demand for digital skills. *Technological Forecasting and Social Change, 162*, 120405 https://doi.org/10.1016/j.techfore.2020.120405. Accessed 13 May 2022.
17. Penuel, W. R. (2022). Implementation and effects of one-to-one computing initiatives: A research synthesis. *Journal of Research on Technology in Education, 38*(3), 329–348. Retrieved 27 June 2022, from https://www.learntechlib.org/p/99387/(2006). Accessed 10 May 2022.
18. UNESCO. (2019). ICT Competency Framework for Teachers. [online]. Available at: https://en.unesco.org/themes/ict-education/competency-framework-teachers. Accessed 17 June 2022.
19. Hawkins, R., Trucano, M., Cobo, R., Juan, T., Alex, S., & Inaki, A. (2021*). Reimagining Human Connections: Technology and Innovation in Education at the World Bank (English).* World Bank Group. http://documents.worldbank.org/curated/en/829491606860379513/Reimagining-Human-Connections-Technology-and-Innovation-in-Education-at-the-World-Bank. Accessed 17 June 2022.
20. Bojović, Ž., Bojović, P. D., Vujošević, D., & Šuh, J. (2020). Education in times of crisis: Rapid transition to distance learning. *Computer Applications in Engineering Education, 28*(6), 1467–1489. https://doi.org/10.1002/cae.22318
21. Beloin, C. A. (2018). *A Study of Customer Relationship Management and Undergraduate Degree Seeking Student Retention.* Thesis, Concordia University, St. Paul. https://digitalcommons.csp.edu/cup_commons_grad_edd/158. Accessed 10 August 2022.
22. Yehya, F. M. (2021). Promising Digital Schools: An Essential Need for an Educational Revolution. *Pedagogical Research, 6*(3), em0099. https://doi.org/10.29333/pr/11061
23. GIGA. (2020). Available at: https://gigaconnect.org/. Accessed 17 June. 2022.
24. International Telecommunication Union (ITU). (2019). The Economic Contribution of Broadband, Digitization and ICT Regulation. Econometric model for the Americas. [online]. Available at: https://www.itu.int/pub/D-PREF-EF.BDT_AM-2019. Accessed 17 June 2022
25. Navaridas-Nalda, M., Clavel-San Emeterio, R., Fernández, O., & Arias-Oliva, M. (2020). The strategic influence of school principal leadership in the digital transformation of schools. *Computer in Human Behavior*, 112106481. https://doi.org/10.1016/j.chb.2020.106481. Accessed 12 May 2022.
26. Trucano, M. (2016). SABER-ICT Framework Paper for Policy Analysis: Documenting national educational technology policies around the world and their evolution over time. World Bank Education, Technology & Innovation: SABER-ICT Technical Paper Series (#01). The World Bank. https://documents1.worldbank.org/curated/en/650911487330772455/pdf/SABER-ICT-Framework-Paper-for-Policy-Analysis-Documenting-National-Educational-Technology-Policies-Around-the-World-and-their-Evolution-Over-Time.pdf. Accessed 17 May 2022.
27. Alghamdi, J., & Holland, C. (2020). A comparative analysis of policies, strategies, and programs for information and communication technology integration in education in the Kingdom of Saudi Arabia and the Republic of Ireland. *Educational Information Technology 25*, 4721–4745. https://doi.org/10.1007/s10639-020-10169-5.

28. Alkahtani, A. (2017). The challenges facing the integration of ICT in teaching in Saudi secondary schools.
29. Alenezi, A. (2016). Technology leadership in Saudi schools. *Journal of Education and Information Technologies, 22,* 1121–1132. https://doi.org/10.1007/s10639-016-9477-x. Accessed 10 August 2022.
30. International Telecommunication Union ITU. (2022). Digital Development Dashboard. Available at: https://www.itu.int/en/ITU-D/Statistics/Dashboards/Pages/Digital-Development.aspx. Accessed 17 June 2022.
31. CITC. Performance Indicators. (2022). Available at: https://www.citc.gov.sa/en/Pages/default.aspx. Accessed 10 Aug 2022.
32. Kozma, R. (2008). ICT, Education Reform, and Economic Growth: A Conceptual Framework. http://download.intel.com/education/EvidenceOfImpact/Kozma_ICT_Framework.pdf. Accessed 12 June 2022.
33. Albugami, S., & Ahmed, V. (2015). Success factors for ICT implementation in Saudi secondary schools: From the perspective of ICT directors, head teachers, teachers, and students. *International Journal of Education and Development using Information Communication Technology (IJEDICT), 11*(1), 36–54.
34. Online Learning Consortium OLC. (2021). K-12 Online Learning in Saudi Arabia. Available at: https://onlinelearningconsortium.org/read/k12onlinelearning-sa/. Accessed 13 May 2022.
35. Ministry of Education. (2020). The Saudi MOE: Leading Efforts to Combat Coronavirus Pandemic (COVID 19). https://iite.unesco.org/wp-content/uploads/2020/10/The-Saudi-MOE-Leading-Efforts-to-Combat-Coronavirus-Pandemic-COVID-19-pdf. Accessed 12 May 2022.
36. Ministry of Economy and Planning, MoEP, Kingdom of Saudi Arabia. (2015). The 10th National Development Plan (2015–2019). http://www.nationalplanningcycles.org/sites/default/files/planning_cycle_repository/saudi_arabia/10th-development-plan-.pdf. Accessed 12 May 2022.
37. Kingdom of Saudi Arabia. (2016). National Transformation Programme (NTP) 2020. https://vision2030.gov.sa/sites/default/files/NTP_En.pdf. Accessed 16 June 2022.
38. Alshehri, Y. A., Mordhah, N., Alsibiani, S., Alsobhi, S., & Alnazzawi, N. (2020). How the regular teaching converted to fully online teaching in Saudi Arabia during the Coronavirus COVID-19. *Creative Education, 11*(7), 985–996. https://doi.org/10.4236/ce.2020.117071
39. Barfoot, R. (2020). Saudi Arabia: COVID-19 education: On-going issues affecting the education sector across the GCC. *Mondaq: Connecting Knowledge and People.* https://www.mondaq.com/saudiarabia/operational-impacts-and-strategy/947066/covid-19-education-ongoing-issues-affecting-the-education-sector-across-the-GCC
40. World Bank. (n.d.). How countries are using edtech (including online learning, radio, television, texting) to support access to remote learning during the COVID-19 pandemic. https://www.worldbank.org/en/topic/edutech/brief/how-countries-are-using-edtech-to-support-remote-learning-during-the-covid-19-pandemic. Accessed 20 August 2022.
41. Affouneh, S., Salha, S., & Khlaif, Z. N. (2020). Designing quality e-Learning environments for emergency remote teaching in Coronavirus crisis. *Interdisciplinary Journal Virtual Learning Medical Science, 2*(11), 135–147.
42. OECD. (2020). *How the COVID-19 pandemic is changing Education: A perspective from Saudi Arabia.* Available at: https://www.oecd.org/education/How-coronavirus-covid-19-pandemic-changing-education-Saudi-Arabia.pdf. Accessed 10 May 2022.
43. World Bank. (2021). Saudi Arabia's Digital and Distance Education: Experiences from the COVID-19 Pandemic and Opportunities for Educational Improvement: Available at: https://www.worldbank.org/en/country/saudiarabia/publication/saudi-arabia-s-digital-and-distance-education-experiences-from-the-covid-19-pandemic-and-opportunities-for-education al-i. Accessed 20 August 2022.
44. UNESCO. (2022). National distance learning programmes in response to the COVID-19 education disruption. Case study of the Kingdom of Saudi Arabia. Available at: https://unesdoc.unesco.org/ark:/48223/pf0000381533. Accessed 17 June 2022.

45. Saudi Ministry of Education. (2020). The statistics of distance education in Saudi Arabia tell an on-going success story in #Madrasati_platform and #iEN_channels (Twitter). Retrieved from https://twitter.com/tc_mohe/status/1330968403780177921?s=20. Accessed 10 May 2022.
46. Han, J., Kim, K. H., Rhee, W., & Cho, Y. H. (2021). Learning analytics dashboards for adaptive support in face-to-face collaborative argumentation. *Computers and Education, 163*, 104041.
47. Karaoglan Yilmaz, F. G., & Yilmaz, R. (2020). Learning analytics as a metacognitive tool to influence learner transactional distance and motivation in online learning environments. *Innovations in Education and Teaching International, 58*, 575–585.
48. Romero, C., & Ventura, S. (2020). Educational data mining and learning analytics: An updated survey. *WIREs Data Mining and Knowledge Discovery, 10*, e1355.
49. Majumdar, R., Akçapınar, A., Akçapınar, G., Flanagan, B., & Ogata, H. (2019). Learning analytics dashboard towards evidence-based education. In *Proceedings of the 9th International Conference on Learning Analytics and Knowledge, Society for Learning Analytics Research (SoLAR)*, pp. 1–6. Irvine, CA, 4–8 May 2019.
50. ITG. (2018). Noor Educational Management system contributes in Saudi schools renaissance. Available at: https://www.itgsolutions.com/noor-education-the-e-learning-solution-that-contributes-in-saudi-schools-renaissance/. Accessed 17 June 2022.
51. Online Learning Consortium OLC. (2020). The State of Online Learning in the Kingdom of Saudi Arabia: K12. Retrieved from https://olc-wordpress-assets.s3.amazonaws.com/uploads/2020/10/v2.3.2-K-12-Report_PUBLICATION.pdf. Accessed 13 May 2022.
52. Security Scorecard. Education Cybersecurity Report. (2018). Available at: https://explore.securityscorecard.com/rs/797-BFK-857/images/SSC-EducationReport-2018.pdf. Accessed 18 May 2022.
53. Tsai, Y.-S., Moreno-Marcos, P. M., Tammets, K., Kollom, K., & Gašević, D. (2018). Sheila policy framework: Informing institutional strategies and policy processes of learning analytics. In *Proceedings of the 8th International Conference on Learning Analytics and Knowledge*, pp. 320–329, Springer.

Chapter 3
New Architecture to Facilitate the Expansion of e-Government

Shreekanth M. Prabhu and M. Raja

Abstract There is an increasing need to look at e-Government as an instrumentality to usher in good governance in a human-centric manner. Conventional e-Governance solutions however focus majorly on automation of operational processes and efficient delivery of services. As far as software architecture goes these implementations typically leverage frameworks and models which are developed for managing large enterprises with complex organizational structures and a diverse customer/consumer base. Further, software architecture over the years has evolved to leverage advances in Social, Mobile, Analytics, Cloud, and agile methodologies. In this paper, we look at how we can make use of an overriding architectural framework that looks at e-Governance holistically by appropriately modelling the people, communities, and societies as well as geospatial and temporal contexts. The setting of goals for Governments can be based on rich data and the change programs better deliberated and defined. Connections between Government and citizens can be direct whether it is in the transfer of benefits or solicitation of feedback. Monitoring of events and roll-out of policies can happen in a near real-time manner. With new and superior architecture, e-Government solutions can lead to high-quality public policy intelligence that not only better inform Governments, but also guide them to formulate more effective governance policies and strategies. We use the Indian context to elucidate our approach. Case studies on effective management of Roads in Bengaluru and on decentralized economic development are covered.

Keywords e-Government · Good Governance · Artha Shastra · Strategic Pivot · Architecture · Ontology · Zachman framework · Tantra framework · Capacity-Capability-Opportunity-Potential Model (CCOPM)

S. M. Prabhu (✉) · M. Raja
Department of Computer Science and Engineering, CMR Institute of Technology, Bengaluru, India
e-mail: shreekanth.p@cmrit.ac.in

M. Raja
e-mail: raja.m@cmrit.ac.in

© The Author(s), under exclusive license to Springer Nature Switzerland AG 2023   55
C. Gaie and M. Mehta (eds.), *Recent Advances in Data and Algorithms for e-Government*,
Artificial Intelligence-Enhanced Software and Systems Engineering 5,
https://doi.org/10.1007/978-3-031-22408-9_3

3.1 Introduction

E-Governance Architectures in many cases suffer from weaknesses that are common to IT systems in general. That is the tendency to automate existing processes without adequate cognizance of opportunities (what stakeholders really would like), potentialities (what all is possible compared to what is currently available), and knock-on effects. On the positive side, new processes can change the behaviour of people and make them more receptive to newer technologies. On the negative side, whenever automated processes come into motion there may be friction and disruption. Thus, *opportunities from stakeholders' perspectives, potentialities,* and *knock-on effects* triad can act as a guidepost for the expansion of the *scope of e-Government*. As in any architectural exercise, *context* plays a very critical role.

The *scope of e-Government* implementations can be described along the four pillars namely (i) *e-Services*: Efficient provision of G2C (Government to Citizen) services through automation and business process re-engineering; (ii) *e-Management*: Management of back-end processes of Governance using automation; (iii) *e-Democracy*: Using electronic means such as e-voting, online petitions, and referendums to further democracy using electronic means; and (iv) *e-Participation*: Participation of Citizens in the process of Governance through suggestions, complaints, and inputs for policy formulation. Here e-Services increase the transparency of governance, improvement in citizen-government relationships, and better revenue for the Government. According to research by Al-Kibsi et al. [1], "the real value of e-Government derives less from simply placing public services online than from the ability to force an agency to rethink, reorganize, and streamline their delivery before doing so". In India, Karnataka Government's 'Sakala' [2] initiative is one such where processes were significantly re-engineered before automation to deliver selected services within stipulated time windows. Whereas major focus has remained on e-Services and e-Management, e-Democracy and e-participation are slowly getting traction. UN Department of Economics and Social Affairs has designed an e-Participation Index (EPI) [3] which is based on: (i) e-information—availability of online information; (ii) e-consultation—online public consultations, and (iii) e-decision-making—directly involving citizens in decision processes. Access to social media and online interfaces to raise grievances as well as the right to information/freedom of information legislation has empowered citizens.

The *opportunities* for bettering e-Governance should align with one or more stakeholder perspectives and enrich the Government-Stakeholder interactions in the process. The *stakeholders* of the current e-Government solutions are Citizens, Businesses, other Government bodies, and employees. Thus, e-Government can be viewed through the prism of the following four interactions. *Government-to-Citizen (G2C)* covers the interaction between government bodies and citizens, involving, among other things, the provision of information and services. This relationship is the main focus of e-Government projects involving citizen participation and civic engagement. *Government-to-Business (G2B)* covers the interaction between government and the commercial sector, involving purchasing/procurement and regulation of business activity, through policies, standards, rules and regulations, and the granting of permits and licenses). *Government to Government (G2G)* covers interaction occurring only among government bodies, either internally or externally at local, state, and national levels. *Government to Employee (G2E)* covers all interactions between the government and its employees, involving matters such as salary, superannuation, welfare schemes, and housing.

We can approach the *potentialities* of e-Governments using Maturity Models. According to Abdoullah Fath-Allah et al. [4] who did a comparative survey of more than twenty e-Government Maturity Models, "Generally, all models contain these 5 stages even though they use different nomenclature. A stage related to the availability of the portal in the Web (presence); A stage where the citizens can interact with governments (interaction); A stage where the citizens can transact with governments (transaction); An advanced stage that covers information sharing between agencies (integration) and Optimization/transformation stage". Thus progressing towards the optimization/transformation stage is one way to realize the potential. Another approach to unleash the potential of e-Government is the Conform-Perform-Reform-Transform framework as follows. With '*Conform*' the focus is on process adherence and people orientation. The government needs to actively communicate with people to provide them with timely service/updates and respond to their day-to-day issues as well as calls of distress. With '*Perform*', a culture of execution with a clear focus on robust outcomes and metrics is expected. This calls for changes in processes, people, and culture, to ensure that things indeed get done on the ground in time. With '*Reform*', a focus on change management with good planning, dialogue, negotiation, and sustained communication with stakeholders during change-over is required. For example, GST introduction to replace many different kinds of indirect taxes. Finally, with '*Transform*' new mindsets not only at the Government level but also at the societal level are imperative. This may be pursuing empowerment in the social sector in place of subsidies, profit and higher Efficiency in the public sector, self-reliance in defense and energy sectors with concerted action, and transformation of complex ecosystems such as agriculture where influential lobbies and vested interests abound side by side with families in distress.

Knock-on effects are unanticipated or inadequately anticipated effects of the action of the Government. For example, automating the delivery of the Public Distribution System (PDS) based on biometric IDs presupposes that citizens have recognizable finger-print, PDS outlets have a working set-up and beneficiaries have the ability to physically present themselves. The same in a traditional setting may happen just based on trust. In the same manner, using a common e-Sign system to authenticate beneficiaries/consumers presupposes an identical recording of their names and attributes. Such assumptions can easily fall apart in large countries. On the positive side, citizens as well as businesses/institutions are becoming more and more comfortable transacting digitally. Digital Payments has had unprecedented traction in most geographies. Governments have also seized the initiative to provide digital certificates in place of paper form. The vaccination certificates in digital form were particularly useful during the pandemic.

The *context* of e-Government varies from country to country due to political, economic, social, and cultural considerations. So coming to local contextual considerations and political factors, the role Government plays in a country becomes a very important consideration for the architectural roadmap of e-Government. This depends on the political and economic ideologies/models different governments follow. Some have free markets and some others mixed economies. Broadly the role of the government can be categorized as owner, leader, partner, and change agent. Even though all governments talk about reducing their role, on the ground however they become all the more pervasive. Table 3.1 prescribes how the Government of India should ideally organize its ministries with e-Government as the strategic enabler, for optimal outcome. As the role of the Government moves toward a change agent, the engagement with society needs to be deeper and the architectural approach needs to adapt accordingly.

Next, we identify the drivers for the expansion of e-Government architecture, with discussion hitherto as the back-drop. Firstly when it comes to stakeholders, the four-fold characterization of Citizens, Businesses, Government Bodies, and Employees falls short. Importantly we need to add society as a stakeholder. Society in turn consists of communities that may have a common identity or interests. How Government engages with specific communities is extremely critical. Individuals tend to be highly influenced by the collective action of the community and may no longer act as rational individuals acting in isolation and making informed individual choices. Secondly, the profession/trade/business people engage in or the company they work for may be yet another grouping, the Governments need to be mindful of. Among them, farmers, unionized workers, and groups that do critical societal functions can act in a highly disruptive fashion. Thirdly, in the era of globalization, no country is

Table 3.1 Government's role (prescribed)

Government's role	Ministries in the Government of India
Owner	Home Affairs, Finance, and Corporate Affairs, External Affairs, Defence, Law and Justice, Parliamentary Affairs, Personnel, Public Grievance and Pensions, Commerce & Industry, Statistics and Program Implementation
Leader	Gas, Chemicals & Fertilizers, Power, Coal, Renewable Energy, Mines, Environment, Water Resources and rejuvenation of rivers, Atomic Energy. Urban Development, Rural Development, Housing, and Urban Poverty Alleviation., Health and Family Welfare, Panchayat Raj, Development of North-East Regions, Human Resources, Skills Development & entrepreneurship, Drinking Water and Sanitation
Partner	Agriculture & Farmers Welfare, Food Processing Industries, Food & Public Distribution. Road Transport & Highways, Waterways Shipping, Railways, Civil Aviation, MSME, Heavy Industry, Electronics & Information Technology, Steel, Textiles, Science & Technology, Earth Sciences, Department of Space
Change Agent	Tribal Affairs, Women and Child Development, Social Justice and Empowerment, Minority Affairs, Sports, Youth Affairs, Culture, Tourism, Consumer Affairs

protected from foreign influence. This may be in the form of foreign-funded NGOs, religious institutions, or charities that do another country's bidding. The media itself may have ideological persuasion and act as a spoilsport. All these need to be tackled with high-quality information that can give a holistic view. Secondly, when it comes to the scope division of e-Governance in the form of services, management, democracy and participation, and approach to e-Governance using Conform-Perform-Reform-Transform vocabulary, we need to expand the scope to include e-Transformation which is defined as "Qualitative Transformation of Governance in general due to adoption of e-Governance" as the fifth pillar.

Now looking at e-Government Architecture as a whole the following considerations become important.

Firstly when a common global vocabulary for e-Government gets adopted there are unstated and implicit assumptions about value systems, culture, and society resulting in lost opportunities due to a lack of leverage of local ethos and heritage. Thus, there is a very important need to root the e-Government solutions in the local milieu, which is the first premise we have followed in this paper. The corollary of this premise is instead of a mere civic model we need to use a civilizational model to understand society and instead of an inward and operational focus, an outward focus that reaches out to society is needed.

Secondly, e-Government is perceived in a variety of ways by decision-makers. Some look at it simply as an automation enabler without fundamental changes to governance. Some look it as a commodity to be bought from vendors and assembled. Thus the second premise of this paper is that the e-Government can act as an instrument of strategic reinvention. Finally, when it comes to e-Governance research, it is best done by anchoring it to one national/regional context. In this paper, we have chosen the Indian context. However, the approach is fairly generic and can be used globally.

The rest of this paper is as follows. In Sect. 3.2, *'Related Work'*, we cover the state of the art in e-Government Architecture. In Sect. 3.3, *'Good Governance'* we delve into the concept of good governance and propose a good governance model, using the Indian context as the backdrop. In Sect. 3.4, *Reimagining e-Government* we address key ideas for the expansion of e-Government. This is followed by Sect. 3.5, *Evolving e-Government Architecture* which describes how using ontology-based Tantra Framework an overarching e-Government Software Architecture can be realized that can prepare us for emerging opportunities and challenges. In Sect. 3.6, *Case Studies/Use Cases,* we look at a Case Study about effectively managing roads in Bangalore and another Use Case that illustrates a novel Capacity-Capability-Opportunity-Potential Model (CCOPM) that takes a strategic approach to decentralized economic development. Section 3.7, Discussions summarizes takeaways from this paper. Section 3.8, *Conclusions* concludes the paper.

3.2   Related Work

Malodia et al. [5] envisage the future of the e-Government using an integrated conceptual framework. They look at empowered citizenship, hyper-integrated network, and
evolutionary system architecture as the important dimensions to conceptualize e-
Government. To define their framework they make use of literature from information technology, public administration as well as business management. They identify
citizen orientation, channel orientation, and technology orientation as antecedents.
Here channel orientation pertains to working with intermediaries. They identify the
digital divide, economic growth, and political stability as well as shared understanding across government departments, and perceived privacy (willingness of citizens to share information in a trusting manner) as moderating conditions in the
e-Government journey. Further, they list cost advantage, time advantage, and efficiency as tangible outcomes and citizen satisfaction and trust in the government as
intangible outcomes. The work draws on insights from hundreds of interviews.

Traditionally Enterprise Architecture is used for e-Government most well-known
among them is FEA (Federal Enterprise Architecture) used by US Government
which was initiated in 1999 and evolved steadily since then. Zheng and Zheng [6]
in the decade-old study of e-government Enterprise Architecture research in China
conclude that most research had focused on frameworks and methodologies but
not adequately on EA policies and actors, principles and standards, implementation
and governance. They also see a lot of opportunity in comparative studies across
countries.

Helali et al. [7] do a study of e-Government architectures in many different countries such as Jordan and Canada to lay out a vision for e-Government in Tunisia.
They identify some of the following as key attributes for e-Government architecture: Interoperability and Integration, Security and Trust, Flexibility, Compatibility,
Traceability, Transparency, Mobility, and Responsibility. Responsibility is defined
as every operation executed on the system must be attributed to a unique identified
legal personality. This legal personality is legally responsible for the execution of the
considered operation, and for all the consequences made public and certified.

Sedek and Omar [8] do a study of e-Government Architecture with a focus on
interoperability and dwell on Technical, Conceptual, Sematic, Dynamic, and Pragmatic interoperability. Interoperability is addressed by advanced architectures and
technologies. Al-Khanjari et al. [9] propose the use of service-oriented architecture for e-Government. Cellary et al. [10] propose the use of Cloud Computing and
Service-oriented Architecture.

Rombach and Steffens [11] in their 2009 paper outline how e-Government has
evolved from one that has an internal focus to improving the living conditions of
citizens and a favourable climate for business. They apply Gartner's hype cycle to e-
Government. They describe the maturity of e-Government using 4 stages: Presence,
Interaction, Transaction, and Transformation. They advocate that e-Government

should proceed on four dimensions: Strategy, Processes & Organization, Technology, and Project & Change Management. They categorize future challenges of e-Government as policy and political challenges and engineering challenges.

Subhajit Basu [12] listed the factors that can introduce risks while implementing e-Government solutions as follows: Political stability (democracy or dictatorial regime), Adequate legal framework, Level of trust in government (perception of service levels), The importance of government identity (fragmentation or integration), Economic structure (education, agriculture, industry or service), Government structure (centralized or decentralized), Different levels of maturity (the weakest part of the chain determines speed) and Constituent demand (push or pull). He lists many factors that impede the e-government environment in developing countries. Among them, Institutional Weakness with symptoms of insufficient planning and unclear objectives resulting in inadequately designed systems continues to haunt developing countries. He refers to OECD guidelines for collecting personal information. These indeed are thought out and need to be scrupulously followed.

Nanayakkara [13] discusses Enterprise Architecture in the context of e-Government in Sri Lanka. They compare the use of SOA with EA based on TOGAF. They emphasize the importance of interoperability at three levels: Technical (systems can exchange data), Semantic (Systems can exchange meaningful data), and Organizational (systems can participate in multi-organizational business processes). Mose and Kimani [14] discuss a proposal for e-Government architecture in Kenya consisting of technology, information, and business/functional. e-government and presentation layers. An exemplary characteristic of the proposed system is that it does not ask for information the government already has.

Tanzania has proposed a relatively modern e-Government Architecture [15] that has a portal with hooks to social media and layered architecture that makes use of SOA. The architecture consists of 9 layers: (i) Consumer Layer that provides the building blocks required to deliver ICT-enabled services and data to end-users; (ii) Business Process Layer enables the realization of business processes by aggregating loosely coupled services; (iii) Services Layer; (iv) Service Component Layer; (v) Operational Systems Layer that provides the infrastructure resource required to run the service components; (vi) Information Layer that includes key architectural considerations pertaining to data architecture (vii) Quality of Service Layer that provides the means of ensuring that the SOA solution meets the non-functional requirements; (viii) Integration Layer and (ix) Governance Layer that governs the entire life cycle of services and their quality. Kumar et al. [16] discuss the central architecture framework for e-Governance System in India where they cover National e-Government Program with many mission mode projects. The state of art as of the year 2015 was about how having a common network that can coordinate multiple projects operating at multiple levels of Government using a strong communication infrastructure.

In recent years Indian Government has come up with India Enterprise Architecture Framework *(IndEA)* [17] to drive its e-Government Program. '*IndEA*' prescribes a plurality of reference models related to performance, business, data, application, technology, integration, security, and governance. 'IndEA' is based on TOGAF architecture. The Government of Andhra Pradesh has come up with e-Pragati [18] scheme which aims to serve as one portal for all government services. Satyanarayana [19] discusses e-Government architecture and implementation in different countries and elaborates on the portal. The thumb rules he mentions are: (i) Focus on Government broadly and not narrowly on e-Government; (ii) Focus on citizens and not on computers; (iii) Focus on transforming the process and not translating the process; (iv) Focus on Software and not Hardware; (v) Plan the Pilot first and not the Roll-out and (vi) Focus on People and not on Systems.

If we look at all these architectural exercises they have a view of governance either as an operation/process or as a service and not as a strategic intervention. The strategic interlock if any is extraneous to the architecture. Understanding of the society that government serves is based on a civic model instead of a social, civilizational model. Political, economic, and cultural factors are also not factored in adequate measure.

If we look at all these architectural exercises they have a view of governance either as an operation/process or as a service and not as a strategic intervention. The strategic interlock if any is extraneous to the architecture. Understanding of the society that government serves is based on a civic model instead of a social, civilizational model. Political, economic, and cultural factors are also not factored in adequate measure.

3.3   Good Governance

The first premise of this paper is for e-Government to be effective it should be rooted in local ethos. That exercise we undertake in the Indian context, by discussing "good governance" as per the ancient Indian tradition. The ancient Indian literature refers to this as a broad topic of Raj Dharma or Duties of King. Discussion on Raj-Dhrama appears in the Shanti Parva [20, 21], a treatise on duties of a king and his government, dharma (laws and rules), proper governance, rights, and justice and describes how these create prosperity. Kaultilya's Arthashastra [22] dated around 230 BCE or earlier is another important text. Bosworth [23] discusses the historical setting of Megasthenes' Indica. Megasthenes was a Greek visitor to India during the times of King Chandragupta Maurya, who had Kautilya as the prime adviser. Megasthenes also refers to other forms of democracy named 'Jana Padas' that were self-organized communities. Even today as it was thousands of years ago, every village in India has a Panchayat (originally meant 5 wise persons) to resolve issues of common concern. Will Durant [24] who did a detailed study of India during British rule found conditions in native provinces ruled by Hindu Kings better than in British provinces. In his words "There are riots between Moslems and Hindus in India. But only in British provinces; strange to say they are rare in native states".

Post-independence, India however chose to ignore that legacy rooted in Indian tradition and pursued a model of governance that was primarily derived from British rule and alien to India. Then they chose to pursue a contentious social compact that was based on Marxist/Socialist ideology instead of a spirit of 'live and let live' rooted in Indian ethos. On top of that, they embraced a peculiar brand of secularism which meant appeasement of communities that are better organized and institutionalized. Added to the mix were populist policies as well as crony capitalism. Even the fundamental rights regime incorporated in the constitution turned out to be highly litigious. Overall India despite phenomenal progress in many indicators continues to face chaos and strife and has underperformed compared to its potential.

Circling back to the present times, our definition of good governance is that addresses stakeholder interests while being mindful of national interest and sustainable outcomes. This is in line with Kautilya's description of the duties of the King where he said "The happiness of the king is in the happiness of the people, his welfare is in the welfare of the people. The welfare of the king lies not in what he desires, but what his subjects desire". Panel of Experts [25] revisited the rich heritage documents in Arthashastra and noted the following learnings as worth emulating by present-day rulers and governments. Hence here we revisit what is good governance as per Indian tradition and look for insights that can guide us. It is plausible that every nation may have similar accounts to share that reflect some form of disjunction or other in their system of governance.

Kautilya states, "An ideal King is one who has the highest qualities of leadership, intellect, energy, and personal attributes and behaves like a sage monarch or *Rajarishi*." Among other things, "a Rajarishi is one who is ever active in promoting the *yogakshema* of the people and who endears himself to his people by enriching them and doing 'good' to them." *Yoga* is explained as the successful accomplishment of an objective and *Kshema* is the peaceful enjoyment of prosperity. The King has no individuality. His duties merged into his personality. He himself was one of the organs of the state, albeit the most important organ. King should have the foresight to avail himself of opportunities by choosing the right time, place, and type of action. He should know how to govern in normal times and in times of crisis. He should know when to fight and when to make peace, when to lie in wait when to observe treaties and when to strike at an enemy's weakness. For Kautilya, *artha* (wealth and larger objectives) followed *dharma* (righteousness). Kautilya stressed that governance should be discharged with a sense of pragmatism. Extreme decisions and extreme actions should be avoided. Soft actions (*Sama, Dana*) should precede harsh actions (*Danda*). Hindu Kings have largely followed the precepts in Arthashastra. Three Good Governance imperatives of Hindu Kingdoms were *Surakhsha* (Security), *Samriddhi* (Prosperity), and *Yogakshema* (Well-being). These imperatives are elaborated below.

The first imperative was safety and security (Suraksha). King's primary role was to protect his people by maintaining order and by preventing attacks by other kings. There was a lot of importance given to dispensing justice. A judge who gave a wrong judgment was subject to penalties. People had easy access to State's high officials and could get their grievances addressed. Tax collectors who harassed people were dealt with strictly. Auditors were dealt with severely if they fail in their duties and for their omissions and commissions which caused losses to the State. King himself was available to meet people on stipulated days/times.

The second imperative is prosperity (Samriddhi). Free trade and entrepreneurism were encouraged. Agriculture, animal farming, and commerce were three pillars of the economy. Kautilya believed that agriculture was the basis for any economy. He insisted on food security and the need to maintain buffer stock, and good treatment of the labour force and was concerned about preventing the misappropriation of state funds. He also advocated just price for products which was based on a reasonable profit margin of 5–10%, demand, and supply. Whenever there was a glut he recommended that the state purchase extra stock and centralize the sale of that commodity at a fixed price. He suggested that only income above fair profit be taxed. He encouraged foreign trade and allowed higher profit margins for imports considering the risks involved. There was a clear-cut role for the state in seeking foreign markets. In summary, unlike in socialistic India, wealth and profit were not anathemas. There was a realization that the wealth of the state depended on the wealth of the people. At the same time, profit had to be fair, the price had to be just and any inter-relationships had to be balanced. The overall model even in ancient India was that of a mixed economy. However, the public sector units were expected to be profitable and the officials who caused losses were penalized severely and those who earned profits were rewarded. There was a strong focus on natural resources and building assets. Kings gave high priority to agriculture, water reservoirs, road building, and forest development.

The third imperative was well-being (Yoga Kshema). The state has to be mindful of taking good care of widows and orphans. Certain jobs widows can do from home were allotted to them on priority. There was a sizable role for the community, with things like education and health managed by the community. The relationship between labour and employer was one of mutual dependency and contracts had to be fulfilled. There was encouragement for artisans who operated in guilds and cooperatives. The government did not interfere with community affairs unnecessarily. There was a strong focus on keeping the rural–urban balance. There were checks to prevent congestion in capital cities. A lot of importance was given to developing villages and rural economies in a self-sustaining manner. There was recognition that the prosperity of the state depended on the prosperity of villages in place of the current focus on unbridled urbanization. There was a strong focus on training and discipline. The Kingdoms supported universities such as Takshashila, Nalanda, and many others with grants. Students were taught Science, Philosophy, Ayurveda, Grammar of various languages, Mathematics, Economics, Astrology, Geography, Astronomy, Surgical Science, Agricultural Sciences, Archery, and Ancient and Modern Sciences, Vastu Shastra (Architecture), and Shilpa Shastra (Construction of Buildings and Temples).

In his work, Kautilya cites and critiques scholars who preceded him who belonged to the continuing knowledge tradition. Then there are other later works built on that tradition such as the *Nitisara* (the essence of policies) of Kamandaki [26]. In essence. Kautilya had a pragmatic approach to Economy, Government, and State-craft, compared to the ideologically dogmatic one many modern governments have. Another key takeaway is a holistic and integrated approach to governance which is in contrast to the modern approach that is reductionist.

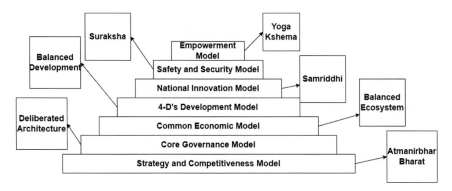

Fig. 3.1 Good governance model

Mahatma Gandhi propounded the concept of "Su-raj (Good Governance)" and according to him "Good governance has the following eight attributes, which link it to its Citizens: (1) Accountable (2) Transparent (3) Responsive (4) Equitable and inclusive (5) Effective and Efficient (6) Follows the rule of law (7) Participatory and (8) Consensus oriented [27]".

Taking these ideas forward, a good governance model [28] is proposed by us shown in Fig. 3.1.

The individual models are elaborated on below.

3.3.1 Strategy and Competitiveness Model

This pertains to the choices India needs to make in order to lead. Daren Acemoglu in his work [29] on "Why nations fail?" talks about two types of economies—extractive and inclusive. An extractive economy primarily seeks rent on natural resources. An inclusive economy is innovation-driven with incentives for the participants. India needs to strike a balance between the two. Being *Atmanirbhar* (self-reliant and self-sufficient) is particularly important.

3.3.2 Core Governance Model

This pertains to core functions of government such as collecting revenue and balancing the budget. The assets of the Government should be used optimally. Wrong projects may be getting funded while projects that can make a difference in day-to-day lives may be ignored. Too many projects may be ongoing without completing any. Here we need to make use of public policy intelligence that takes a holistic view. The governance architecture needs to be deliberated well.

3.3.3   Common Economic Model

This pertains to defining the Public, Private, Foreign, and Cooperative sectors to achieve economic prosperity. The economies of the world have moved from being driven by factors of production (land, labour, and capital) to agglomerative and conglomerate (clusters of industries) to an economy that is networked and distributed. India can leverage its scale to enhance networking and ease distribution. The intent here is a vibrant yet balanced ecosystem.

3.3.4   4 D's Development Model

India needs to work out the 4D's of development—distribution, differentiation, depth, and disintermediation. The use of digital and social platforms along with high-quality information on societal needs and capabilities can help here. The goal here is balanced development.

3.3.5   National Innovation Model

This pertains to developing new technologies, skills, practices, business models, and platforms. In addition, choosing the right strategies for capacity creation and utilization, capability enhancement, and converting potential to opportunities is critical. Innovation should be the engine for wealth creation.

3.3.6   Safety and Security Model

Protecting citizens' lives, property, and liberty from all kinds of attacks and establishing a security presence in physical and cyber-space is a daunting task that needs a way to capture and connect all kinds of information.

3.3.7   Empowerment Model

People need access to housing, good nutrition, health, education, and employment opportunities. Government should strive to meet higher-order needs leaving basic needs to self-organized and self-sufficient communities.

In the ultimate analysis, a good governance model should transform the lives of citizens. Toward that purpose, we make use of Maslow's Motivational Hierarchy of

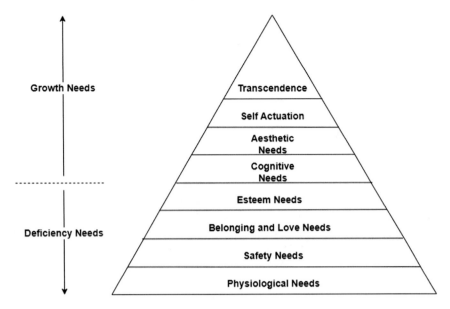

Fig. 3.2 Expanded Maslow's motivation model

needs as a guiding framework to think of desired outcomes. Figure 3.2, shows the expanded hierarchy of needs [30] in tune with the present.

It is our expectation that as the foundational layers (or the base) of the Good Governance Model get stronger they make a greater difference to higher layers significantly empowering people and in the process meeting the higher needs of Maslow's Motivational Model.

In summary, in this section, we drew on the foundational ideas from ancient Indian tradition as to what constitutes good governance, and then we proposed a contemporary good governance model linked to the 3 main imperatives—Suraksha (Safety and Security), Samriddhi (Prosperity) and Yogakshema (Well-being of people). Implicit in our exercise is the belief that the prevalent governance models in countries such as India have underperformed.

3.4 Reimagining e-Government

The second premise of this paper is for e-Government to be an instrument of strategic reinvention. In other words, e-Government should not be seen merely as a driver of efficiency but as an instrumentality for strategic reinvention. We conceptualize reinvention using the following strategic map in Fig. 3.3. This approach is based on the mental model consisting of 3 questions: Where are we now and where do we want to go? How do we get there? And how do we further the journey? Here we make use of the 3-phase redefine-refine-lead framework. In "Redefine Phase",

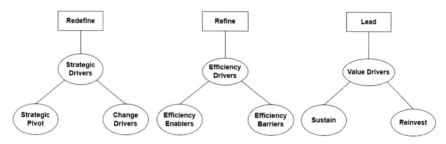

Fig. 3.3 Strategic map

we identify strategic drivers, and change drivers and dwell on strategic pivots and opportunities associated with them. During the "Refine Phase", we look at efficiency drivers, enablers, and barriers. Finally in the "Lead Phase", we look at value drivers and how value can be sustained and nurtured using appropriate reinvestment. These phases are cyclical. The concept of Strategic Pivots used here is due to Michael Irwin [31].

3.4.1 Redefine Phase

Here we start with strategic drivers to reimagine e-Government. Firstly, we should have an approach that is more pragmatic than based on ideological dogmas. Pragmatism is highly emphasized in Kautilya's Arthashastra. Secondly, when it comes to governance extreme and abrupt actions should be avoided lest there will be negative knock-on effects. This means any change needs to be debated, ruminated in the collective conscience, and only then adopted, but without losing the opportunity. Thirdly, instead of optimal outcomes, we should look at the middle path. Maximally benefitting a given constituency or a cause may lead to undesirable outcomes for other causes and constituencies. In other cases, there may be long-term implications. For example, a sudden switch-over to organic farming was one major cause cited as why the Sri Lankan economy got into trouble. The policy choices thus need to balance global trends and local practices/ground reality carefully. Governance needs a holistic integrated approach that factors economic, political, social, cultural, and civilizational considerations. Suraksha (Security), Samriddhi (Prosperity), and Yogakshema (Well-being) can serve as 3 pillars of good governance We propose the use of the 7-layer Good Governance Model to look at all aspects of governance in a holistic manner. A holistic approach is also needed in temporal and societal dimensions. That means performing longitudinal studies and ecosystem studies to validate strategy/policy choices.

Coming to change drivers, there is a greater focus on being self-reliant and self-sufficient for all nations. This is a departure from the importance of core competency theory where each nation focuses on what it is good at. The other trends are a convergence of the social, physical, economic, and cyber world when it comes to natural security; convergence of trade, frauds, and threats; disruptive foreign influence in the form of NGOs and civil society; pervasive digital inclusion; and globalization of opportunities. Many times using global rankings to drive change can mislead nations. Ideally, each nation should define its own path forward. Before taking up any new project, rich public policy intelligence that assesses the opportunity costs of projects among multiple alternatives should precede any change. Every change needs to be socialized adequately before launching it, as benefitting one section of society may cause disaffection in others. Governments also persist with bad policies for the fear of upsetting people because of a phenomenon called pluralistic ignorance, where everybody overestimates the number of people opposed to the change.

The strategic pivots that we can pursue are focusing on righteous action instead of just process or outcome focus; a multi-scale approach to governance that makes measurements at the level of community, locality, region, and nation; monitoring and mining processes in operation; refined and nuanced management of social information that can accommodate fuzziness and polymorphic processing, as needed.; making use of decentralized approaches to governance with available technology choices.

3.4.2 Refine Phase

Here we look at efficiency drivers, enablers, and barriers. We propose the following efficiency drivers: comprehensively defining objectives and linking them to measurable outcomes; pursuing the right objectives at the right time and in the right order; choosing the right policy interventions; selecting the right beneficiaries to provide relief in the event of disasters—floods, drought, etc.; refining targeting of benefit s by getting the right granularity for applying policies. The families who need it need to get it in good measure. Ensuring fairness in the application of policies and implementation. The efficiency enablers are ensuring monitoring and examining why outlays are not leading to outcomes, transparently communicating to people how taxes paid by them are spent, and delivering services to people within a reasonable time; The efficiency barriers we identify are skewed prioritization, market distortions, and bad behaviour. Any subsidy and mispricing distort markets and if availed by those who are not needy, it makes the matters worse.

3.4.3 Lead Phase

Here society with the right value systems and culture is the biggest driver for good governance. The value system needs to be sustained and nurtured by promoting good behaviour. Similarly, Government needs to do its part. The value drivers we identify are ensuring that all the people pay their part of taxes; inculcating respect for rule of law in society; encouraging people to interact with the Government through complaints, suggestions, and advocacy instead of street protests that cause huge losses to the economy and disrupt lives. In order to sustain the value, the Government needs to be accountable to people, and earn their trust say by communicating and demonstrating that taxes are well-spent. Communication and engagement are all the more important In case of major changes. To continually nurture the value created and to attain sustainable leadership, the Governments need to develop a comprehensive understanding of society and inter-relationships within and only then the mindsets of people can be changed to operate in an empowered fashion. The societal ethos and values such as innate belief in doing one's Dharma (duties and righteous conduct), any kind of debt entailing a sacred obligation to replay and not causing unnecessary violence and annoyance to fellow-human beings and life forms as well as respect for nature and need to keep neighbourhood clean can go a long way to enable a nation to lead.

In summary, in this section, we have looked at the kind of mind-sets required to make strategic choices and requisites for operational efficiency followed by the need to develop mutually respectful and trusting relationships between society and government. These ideas can broaden the architectural landscape of e-Government significantly.

3.5   Evolving e-Government Architecture

In this section, we aim to evolve an e-Government architecture that caters to the good governance model and strategic map discussed earlier. The discussion in this section makes use of the Indian context. Contemporary e-Government Architecture either takes off on Enterprise Architecture such as TOGAF or on IT Architecture such as Service-Oriented Architecture (SOA). Generally, they cater to internal operations to manage processes within the government and provide an interface, point solution, or mobile application. The citizens are viewed using a civic model where they operate as a collection of individuals instead of a social/civilizational/community model that is typical of traditional societies. We propose an approach that models society comprehensively, sets goals, assesses separations/gaps, and designs interventions in a cyclical fashion. The societal model we are using subsumes the civic model. See Fig. 3.4.

The way society is modeled has a strong influence on policy architecture. The current way of modeling society as levels (below and above the poverty level, income levels), categories/classes, and communities (majority/minority) is simplistic. Looking at rich farmers, aggregators, intermediaries, marginal farmers, owner farmers, tenant farmers, and farm workers as one monolithic group called farmers is not right. This leads to a broad-brush approach to policies regarding income tax exemption, loan waivers, scholarships, etc. In the case of quotas and scholarships, policies based on category and community may lead to poor targeting. A lady may belong to a wealthy family but not be taken care of by her children or be dispossessed of her assets. The urban populace may have higher incomes but a

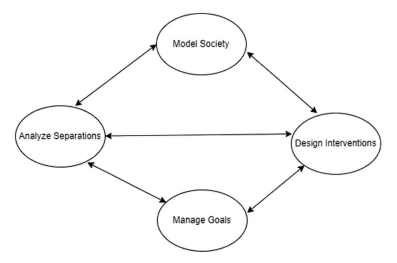

Fig. 3.4 Societal model of governance

poorer quality of life. Law-abiding middle classes may get no benefits as the policies exclude them from getting scholarships, health-care benefits, and loans at reduced rates and they may be taxed the most, creating an asymmetric relationship between them and the government. This is when extending schemes such as education loans more liberally may immensely empower them. Overall we need an approach that is fine-grained and multi-scale. Society is best represented as networks and overlapping communities that transcend any stereotypical premises.

To elevate e-governance to good governance that results in social change, we need an overarching e-government Architecture Framework that can capitalize on strategic opportunities. To that end, we propose an Ontology-based Tantra Framework, which allows for unified modelling of Society, Government Institutions, relationships, and networks as well as underlying phenomena. Tantra framework builds on Zachman's Enterprise Architecture Framework [32] which has aspects and perspectives. The aspects are derived based on English Language Interrogatives namely who, what, where, when, why, and how. The perspectives facilitate the reification process from concept to enumerated construct. We have extended the Zachman Framework which modelled enterprises with a focus on IT systems and arrived at Tantra Framework to model society and Government seamlessly. Each aspect can cater to an instance, sub-set, or set. We make use of the Neo4j Graph database that allows any level/node to connect to any other level/node. The aspects can be used to model fuzzy sets as well with probability as an accompanying attribute. We make use of relator and relationship concepts from Unified Foundational Ontology [33] and incorporate them as additional aspects. Further, we introduce separations as the 9th aspect. The framework can enable sector-specific modelling as well to represent sectors such as Financial and Agriculture. Here Governance processes can be modelled using the Process column and Government Departments can be modelled using the Relator column.

The aspects of the Tantra Framework are described below.

1. People/Households/Communities/Categories (Who)
2. Places/Addresses/Locations/Zones/Geo-spatial Constructs (Where)
3. Assets/Attributes/Entitlements (What)
4. Events/Seasons/Durations/Timelines/Temporal Constructs (When)
5. Processes, Interventions, Phenomenon (How)
6. Objectives, Measures, and Metrics (Why)
7. Relationships between Aspects, Transactions, Affiliations, Social Networks, Ecosystems, Supply Chains, Value Chains, Social Circles (Relationships)
8. Relators/Intermediaries/Institutions (enable relationships and entitlements)
9. Separations (difficulty of establishing a relationship).

Each aspect of the Tantra Framework is reified using perspectives at contextual (named), conceptual (defined), logical (designed), physical (configured), and instantiated levels. Table 3.2, addresses the reification of the People Domain for different roles. Here we make use of the notion of Domain and Roles as described in Codd's seminal paper [34]. Table 3.3 gives indicative examples for other aspects.

If we want change to happen, we need to make use of the framework in a normative mode, going beyond the descriptive mode. We define Tantra Normative Framework in Fig. 3.5. Here, the process of change comprises setting goals, designing interventions, and measuring separations (which in turn leads to recalibrating the goals and interventions as things change) and operates on an information space created using Tantra Framework. Here we propose the Balanced Scorecard Methodology [35] for setting goals, Bartels' Theory of Separations [36] for arriving at separations, and the

Table 3.2 Reification of people domain

Perspective	All people	Citizens	Residents	Aadhaar-Card holders	PAN card holders
Named (identified and contextualized)	All the people within the identified context	People who are citizens	People who are residents	People who have an Aadhaar Card	People who have a PAN Card
Defined (conceptually structured)	What makes one a member of this domain/role	What makes one a member of this domain/role	What makes one a member of this domain/role	What entitles one to an Aadhaar Card	What entitles one to a PAN Card
Logically Designed	Related attributes that map to other aspects	Related attributes that map to other aspects	Related attributes that map to other aspects	Related attributes that map to other aspects	Related attributes that map to other aspects
Configured	Representation in Graph database	Representation in graph database	Representation in graph database	Representation in graph database	Representation in graph database
Instantiated	Instantiate with a unique ID	Instantiate with a unique ID	Instantiate with a unique ID	Instantiate with a unique ID	Instantiate with a unique ID

Table 3.3 Other aspects and examples

Aspect	Examples
Address	Residential address, general address/location, commercial address, institutional address, address for civic amenity
Event	The birth event, education enrolment, employment, marriage, retirement, death, etc.
Asset	House owned, vehicle owned, land owned, business owned, stocks, etc.
Processes	Aadhaar enrolment, voter enrolment, PDS enrolment, birth registration, property registration/lease
Artefacts	Aadhaar Id, Voter ID, Public Distribution Service (PDS) card, Permanent Account Number (PAN) Card, Birth Certificate, Death Certificate, Property Sale/Purchase, Property Transfer/Ownership. Lease
Objectives/Metrics	Citizen satisfaction, poverty level, service turn-around times, employment coverage
Relator	Enrolment agencies, service providers and government departments
Relationships	Familial relationship (marriage), social relationship (community). affiliative relationships (employment), privilege (driving license from a RTO), benefit (PDS entitlement in specific zones), voting privilege (in segments, booths)
Separations	Financial, informational, spatial, temporal, knowledge, social

Fig. 3.5 Tantra normative framework

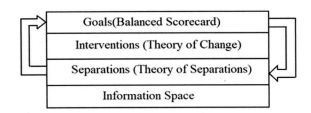

Theory of Change [37] for interventions. The framework however is generic and can incorporate any other methodology.

Table 3.4, gives a list of proposed metrics and indices based on balanced scorecard perspectives.

Further Tantra Framework can be conceptualized as Tantra Digital Governance Framework as shown in Fig. 3.6. Here the information space is heavily discretized and digitized. Tantra Framework is particularly useful to create a digital universe. The framework is modelled along the lines of TM Forum's Framworx [38]. Here we count on information champions belonging to the government, public, social and institutional sectors to help populate the information space. All information goes through a veracity check. The laws, policies, and processes are enabled using the Governance Layer.

The framework is implemented using Neo4J Graph Database which enables highly flexible connectivity that can support multi-relational networks. Figure 3.7 illustrates how the People aspect is represented. Figure 3.8 represents relators who represent Government departments. Figure 3.9 represents geospatial separations faced by people.

Table 3.4 Good governance metrics

Financial perspective	
Capacity utilization index	Ability to utilize/monetize national resources
Productivity	Needs a nuanced approach to define this. Related to capacity utilization
Competitiveness index	Global or internally defined index (checks if we are prioritizing the right sectors)
Revenue capture index	Ability to make revenue for the government due to economic activity
People perspective	
Mobility index	Social, economic, and occupational mobility over time
Citizen satisfaction index	Satisfaction related to various aspects of governance
Community satisfaction index	Satisfaction related to various aspects of governance
Business satisfaction index	Satisfaction related to various aspects of governance as far as G2B interfaces are concerned
Collective action index	Frequency of protests and disruptive actions vis-à-vis advocacy and other methods
Process perspective	
Fairness index	Ability to deliver benefits, services, and opportunities fairly across the society
Leanness index	Re-engineer the process to only essential steps based on causality
Agility index	Track the process for the timeliness
Learning and growth perspective	
Learning index	Learning new behaviours
Social innovation index	Innovation Intensity across society
Diversification index	Structuring of economy among private, public and foreign or any other combination that better prepares the country for future
Networking index	Growth of the economy due to digitally enabled networks peer-to-peer interactions
Ethical perspective	
Ethical behaviour index	Proclivity to behave ethically
Work ethic index	Excellence as an ethical value

Tantra Framework has been applied to a variety of scenarios related to e-Governance. Table 3.5 details different scenarios where Tantra Framework is used and highlights. The papers [39, 40] cover the most recent literature on Tantra Framework.

In this section, we have proposed Tantra Framework a conceptual Framework that is ontology-based. It caters to 9 Aspects and 5 perspectives that enable the holistic, integrated, and seamless representation of information pertaining to e-Governments. The Tantra Framework can be repurposed as a normative framework where goals, separations, and interventions can be modelled. Further, Tantra Framework can be

Fig. 3.6 Tantra digital governance framework

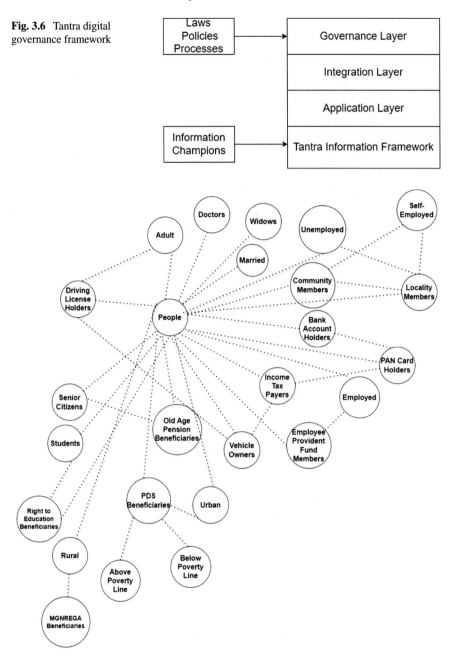

Fig. 3.7 'People' aspect represented in a varied manner

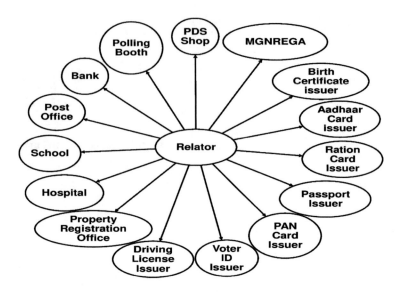

Fig. 3.8 Government and public institutions represented as relators

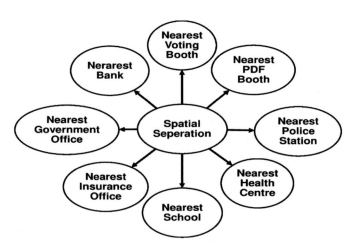

Fig. 3.9 Geospatial separations

expanded to realize a Digital Governance Framework where a plurality of applications can be integrated and governance instrumented. In summary, software Architecture based on the Tantra Framework can facilitate the expansion of e-Government significantly.

Table 3.5 Tantra framework application scenarios

Scenario	Highlights
Identity management	Here we propose federated identity management using the Tantra Framework which can relate multiple forms of identities to each other. Any change in personal information is tracked using events
Revenue capture	This focuses on property tax recovery which is a common problem where data available with the government may be stale and inaccurate and the ability is needed to trace the defaulters. There also may be issues with poorly designed processes that act as blockers
Intelligent social banking	Here rich information is used to enable peer-to-peer lending networks with a role for banks and insurance companies to achieve financial inclusion
Balanced development	Here Tantra Framework is used to track inequalities while pursuing development and take corrective actions
Balanced growth	Here we reimagine the classic balanced growth model to drive the Indian Economy to new heights
Electoral democracy	Here we propose leveraging Tantra Framework to ensure complete and correct electoral rolls on an ongoing basis to ensure voting by all sections in particular middle classes
Social sector	Here we analyze change theories and propose a flexible social benefits plan instead of subsidies enabled by the Tantra Framework
Agricultural ecosystem	Here we repurpose Tantra Framework to address the complex agricultural ecosystem and corresponding information ecosystem that spans biological, social, economic, business, industrial, and welfare ecosystems
Validation of social information using entropy	Here available social information is clustered and networked and the state of completeness/compactness of information space is assessed using a novel entropy construct, This approach can be used to validate census exercise

3.6 Case Studies/Use Cases

In this section, we look at two case studies/use cases, where Tantra Framework can be applied. The first use case is on Managing Roads in Bengaluru. The second use case is related to de-centralized economic development.

3.6.1 Managing Roads in Bengaluru

Even though Bengaluru/Bangalore in India is home to a large number of MNC Captives and important public institutions the state of roads is generally poor leading to long commute times and accidents as well as pollution. Here we are not only referring to roads but also surrounding civic infrastructure that includes footpaths, street lights and transmission, communication lines, and any other setups, housed on the side of the road. Other issues are potholes on the road, flooding of roads, and debris on the road or on the side. Potholes and minor work are neglected as there are fewer incentives for all involved.

To alleviate congestion civic bodies take up work such as flyovers which add to the inconvenience of commuters. In many cases a large number of projects are ongoing and new projects are started without completing the current ones. Similarly, there are white-topping of roads which again take long durations. Many projects once completed are not finished properly due to a lack of intent and the multiplicity of agencies involved. Net-net the city is always in a "work in progress" state. When certain sections get streamlined other parts do get congested.

Some roads are made up using white-topped concrete which is expensive but long-lasting. However such projects happen at the cost of many other smaller projects. In places where work is done, it does not happen neatly, debris is left over. The roads are not laid end to end. Every work happens in bits and pieces. Some roads have experimented with bicycle lanes without a continuous stretch and adequate road space. Some other roads are reconfigured to accommodate bus lanes, without clarity on effectiveness. Side-by-side construction related to the metro transport network happens adding to the confusion. Sheer diversity of vehicles is astounding. Some vehicles are very small and some very huge, while some very old and some very modern, all sharing the same road space. Add to this, the behaviour of some drivers can be bad. Topography is ill-thought-out compelling some drivers to drive on opposite sides. In addition to roads, more critically the footpaths have proved to be highly unsafe. There are also other considerations such as safety and livelihood. Some drivers driving school buses and food delivery vehicles are constantly racing against time.

The same city was called Garden City and a paradise for retirees. Even now the campuses of Public Sector Units are maintained with good roads and street lighting. It is only after the IT boom and large-scale migrations into the city the quality of the roads has deteriorated. There is also a lack of aesthetics and a sense of pride Table 3.6, illustrates the modelling of use case in the Tantra Framework. Table 3.7 describes the analysis of the use case.

We propose a few interesting route metrics here: Geodesic multiple is the actual distance versus shortest distance based on Geotag coordinates. If this ratio is very high, there may be an opportunity to propose intermediate roads by removing any blockers. Spatial Congestion multiple is the time taken to travel a stretch divided by the time taken to travel any other stretch of the same length in the zone of interest. Temporal congestion is the time taken to traverse a stretch during peak time over

Table 3.6 Modelling Bengaluru roads

Aspect	Description
People (Who)	Community in a locality, employees in a company, commuters on specific routes, street vendors, school children, senior children, women, companies, schools, neighbourhoods, resident association, drivers, owner drivers
Localities (Where)	Central zone, outskirts, civic wards. Assembly constituencies, districts, Bangalore north, south, west and east
Time zones (When)	Morning peak, evening peak hours, night hours, industry commute hours, school hours, hospital hours, fault reporting event, start of repair event, new project initiation event, new project inauguration, dependent event (such as land acquisition)
Roads (What)	Side roads, arterial roads, high ways, village roads, new roads, footpaths
Vehicles (What)	Autos, SUVs, trucks, bullock cards. buses, cars, school buses, luxury buses, out station buses, two wheelers, tempos, transport vehicles, water tankers
Facilitators (What)	Street lights, signals, CC TVs
Barriers (What)	Potholes, discontinuities, recurring potholes, flooded streets, blocked streets. broken footpaths, intruding/unsafe cables, debris, shops on footpath, construction of new apartments, hurdles to start a project, delay in payment to contractors
Process/Project (How)	Road repair process, road improvement project, laying new roads, filling potholes, flyover, underpass, junction widening, concreting the road, white topping road, laying new roads, projects by utility companies
Ratings/Metrics (Why)	Mean time to report pothole, mean time to repair, civic quality rating, speed/km, citizen satisfaction index, turnaround time for projects, project closure rate, project concurrency rate, project quality rating. sense of affinity and pride with city, citizen cooperation index, senior-citizen safety, child safety, tourist safety, opportunity cost of projects. project priority, benefit, duration of projects
Routes (Relationships)	Routes connecting any two localities, stretch between any two points, hubs and spokes, junctions, and roads
Route metrics	Geodesic multiple, spatial congestion multiple, temporal congestion multiple
Junction metrics	Junction delay, junction curvature, junction density
Relators	Police authorities, civic authorities, corporates, MLAs, MPs, BMTC, BMRCL, BESCOM, BWSSB, utility companies (telephone, gas), transport department, ministers, contractors, civic work departments, mobile-app service drivers, bus drivers, school bus drivers, professional drivers
Separations	Disjunctions between bus routes (spatial separation), routes that cause temporal separation due to congestion, Routes that are expensive to travel (financial), informational separation (on projects and plans), capability separation (lack of expertise), social separations (certain areas neglected and lack of avenues to complain to the right people)

Table 3.7 Analysis of Bangalore roads use case

Aspect	Analysis
People communities groups	Work out most affected groups due to congestion, safety, pollution, and general civic amenity deficit and note common and unique concerns
Locales	Analyze the most affected localities using residential and commercial considerations as well as geospatial considerations
Time-zones	Perform day/time-zone wise analysis across locales and people based on their requirements
Roads	Analyze correlation between the type of roads and issues in them
Vehicles	Perform analysis based on vehicle types. Check if oversize or peculiar vehicles cause congestion/accidents
Drivers	Analyse drivers as owners as well as relators. Analyze driver behaviour and impact
Routes	Analyse stretches contributing to congestion/accidents/civic hazards and most vulnerable routes
Projects/Repair work	Analyse best practices and well-performing projects. Look at opportunity costs among all plausible projects
Metrics, relators and separations	Make use of a collection of metrics to understand the root cause. Suggest suitable actions to be done by different agencies in the short, medium, and long term

time taken during other times. The topology of a junction whether it is square or curved is another metric that can affect flow of traffic.

We propose periodic Civic Audits which are enabled by a civic governance portal that captures all the information powered by Tantra Framework. This will require forming committees of citizens that do the audit and report to the government and the government, in turn, communicates its plans. If the Audit is planned every two months with inspections across the city and wide communication is done, a lot can be changed. Some of the projects can be funded by Local Area Development Funds of elected representatives. At any point in time, there should be a prioritized list of projects shared in the public domain. All work from planning to completion should be done transparently with a scope of participation for all citizens. Now when work is happening in the neighbourhood, citizens have no idea why was it initiated and how long it will happen and why was it prioritized. It is common to see white-topping of roads happening side by side with broken footpaths. The latter probably takes a fraction of the money. It also may be a good idea that instead of devolving funds from other taxes, the city should be run by charging civic access charges to both citizens and businesses and spend the money in a transparent manner. Now property tax collected is spent in a very opaque manner. Many times wrong projects are prioritized. Here Tantra Framework can enable high-quality public policy intelligence.

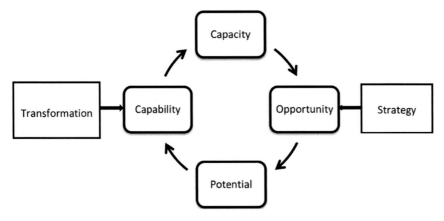

Fig. 3.10 Capacity-Capability-Opportunity-Potential Model

3.6.2 Decentralized Economic Development

One of the reasons for a lot of issues in India is the economic model it has focused on. The model with macro-level centralized planning has resulted in excessive focus on urbanization which has come with its own share of problems. Secondly western model of pursuing productivity and automation to promote development is not suitable for India which is rich in human capacity. Hence the second case we propose is Decentralized Economic Development where each district, community or locale defines and drives its own destiny. To that end, we propose a Capacity-Capability-Opportunity-Potential Model (CCOPM) as illustrated in Fig. 3.10.

Given a district, community, or locale (DCL), there may be the capacity (say human capacity) that needs to be matched with the opportunity to realize a productive outcome. But the capacity identified may not have the required capability. On the other hand, there may be capability say with highly skilled artisans but their numbers may be too small. In this case, we need to transform capability to capacity, while in the earlier case existing capacity may need to be imparted with requisite capability. Once we have the required capacity coupled with the capability it can be matched with a specific opportunity such as creating marketable artwork or craft items.

In other cases, there may be a good number of talented actors but they may not have a suitable opportunity for them to showcase their talent and lead a meaningful life. But if the given DCL has the potential to be a tourist destination, then the talent can be exploited. Alternatively linking them with digital platforms to showcase their talent may be another option. Another potential that can be looked at may be growing fruits and vegetables, terrace farming, developing specialized technical skills, etc.

The opportunity can be as mundane as opening a Petrol Bunk when the numbers that serve a populace are too few. It can even be natural resources such as lakes and ponds or high-quality schools or leisure farms. Maybe certain areas are gifted with natural resources such as minerals. Others may have knowledge related to Ayurveda.

Some may simply serve as Centres for scholarly pursuit. Some districts may have the potential for gaining business from other districts. Starting local language BPOs, selling handicrafts, produce, or preparation that is unique to that district nationwide.

Even in urban locales, setting a planned and self-contained Software City/Academic City which meets all its needs including housing, schools, hospitals, and other amenities can be an approach to follow. This is preferable to just transplanting multiple company offices in a dispersed manner within an unplanned and crowded city. An art museum can promote art. A weekly market/digital platform can enable farmers to present their produce directly to customers. An aggregator/intermediary can help expand markets and even create markets.

Strategy comes when we choose the right opportunities and the right themes. Thus, in a given locale those institutions that do not fit with a given theme may be nudged to move out. Transformation comes in when there is a continued attempt to promote capacity and capability on one hand and potential and opportunity on the other.

In this case, Tantra Framework can help model society across groups of people, locales, and markets by working with relators who may be cooperatives, advocacy, or promotional organization. People are modelled as communities who are associated with a locale where they mutually characterize each other. Here potential is modelled as separation, once bridged it can convert into opportunity. Here Tantra Framework can facilitate the right networks and nurture an ecosystem that can help create prosperity in a sustainable manner. Table 3.8 shows the modelling of the use case using the Tantra Framework.

Traditional economic models look at factors that primarily pertain to deficiency needs in Expanded Maslow's Motivation hierarchy (Fig. 3.2). Here, constraints galore. If we however focus on meeting growth needs such as cognitive, aesthetic, and self-actualization, only our minds can limit us. No wonder Kings of yore took pride in patronizing fine art, knowledge, and craftsmanship.

3.7 Discussions

The goals we set out for ourselves in this paper was to expand e-Government along 5 criteria—scope, opportunities from stakeholders' perspectives, potentialities and context.

When it comes to scope, the proposed architecture is well-suited to support the four pillars of e-Governance namely e-Services, e-Management, e-Democracy, and e-Participation particularly well as integrates social information and information pertaining to processes of governance whether it is objectives, processes, etc. in a unified manner. Further, it becomes easier to arrive at the right interventions and then monitor them in a near-real-time manner by enabling direct connections with citizens. People can be modelled as communities, professional groups, families, networks, micro-networks, and so on. Each such aspect can be linked to other aspects such as location, asset, and event resulting in relationships among the aspects. The

Table 3.8 Modelling the decentralized economic development

Aspect	Description
Capacity (community and locale)	Artisans, technicians, traders, skilled specialists innovative farmers, artists, actors, dancers, musicians, knowledge owners, ready-to-work workforce, sportsmen, mathematicians, scientists, scholars, teachers, wrestlers, traditional skilled artists, communities with traditional knowledge, tilling capacity, capacity of locales, capacity of districts
Potential (Separations)	Natural resources such as minerals, potential for civic amenities—petrol bunk, school, college, hospital, bank, small finance institutions, tourism, agricultural potential, water resources, connectivity
Opportunity	Mining, commerce, tourism, training, knowledge, export, specialty agriculture, BPO, folk art, luxury farming, water body, roads
Capability (Metric/Objective)	Capability of communities, capability of individuals, capability of locales, capability of districts
Ecosystems (Relationships)	Value chain, supply chain, knowledge network, market network, peer-to-peer lending, farming ecosystem
Strategy (Why)	Themes, choices (where to lead, lag and be on parity), natural resource strategy, tourism strategy, skill-based strategy, knowledge-based strategy, craft-based strategy, trade based strategy, innovation-based strategy, traditional knowledge-based strategy, fine-art/folk art based strategy
Transformation (Process)	Capability to capacity, capacity to capability, potential to opportunity, capacity utilization, social/economic mobility, quality of life and environment

performance of government can be analyzed by assessing the relationship between relators (government departments) and citizens at varying granularities. Thus the capability of e-Government solutions can be expanded manifold and analysis can be conducted at micro, meso, and macro levels. The architecture is particularly amenable to multi-scale analysis.

Next, we look at opportunities from stakeholders' perspectives. In Tantra Framework we look at the societal/civilizational model of citizenry which subsumes the civic model. In the civic model, the typical stakeholders are citizens, businesses, other government bodies, and government employees. In the societal model, we take a broad approach and handle concerns of communities, families, professionals, age groups and any social/professional networks belonging to a particular locality for a particular time period in a granularity that can be as fine or as coarse as desired. Further people can be grouped as overlapping clusters. This capability enables e-Government solutions to go beyond simplistic categories and levels and enable governments to formulate policies in a judicious manner and apply them intelligently. In the same manner, we can do eco-system level analysis going beyond individual businesses or industries.

When it comes to potentialities, Tantra Framework can be a valuable companion as e-Government attains higher maturity levels covering presence, interaction, transaction, integration, and optimization/transformation. This journey will not happen without concomitant social change. An alternate way to look at this is using Conform-Perform-Reform-Transform Framework. Here Tantra Framework with its ability to instrument measurements at the societal end as well as at the government end through relators/intermediaries, processes and objectives can achieve this four-fold improvement in a continuous manner.

Next, we look at knock-on effects. To temper any knock-on effects and ensure net-positive changes we make use of a strategic map approach. This encompasses 3 phases names Redefine, Refine and Lead. In the Redefine phase, we have recommended the middle path and incrementalism to temper any knock-on effects. We also look at change drivers and strategic pivots that can enable Governments to make the right choices after assessing opportunity costs. In the Refine Phase, the efficiency drivers, enablers, and barriers are identified. The Governments can take note of these and get their operational processes to be effective till the last mile. Here again, Tantra Framework can be handy. In the Lead Phase, we look at value drivers, here the focus is on social change and promoting middle-class values. Tantra Framework-based architecture enables the collection of social information in a unified manner to support this continuous and cyclical transformation.

Finally, we have captured philosophical underpinnings by revisiting ancient Indian tradition and ethos and corresponding good governance model and proposed an approach and roadmap to realize it.

3.8   Conclusion

As French writer Jean-Baptiste Alphonse Karr wrote "plus ça change, plus c'est la même chose" i.e. "The more things change more they remain the same". Across the world, many Governments have embarked on a journey to transform themselves into e-Governments with huge expectations but challenges remain. So do opportunities that are yet to be tapped. In this paper, we have reimagined e-Government using the Good Governance perspective where Society is the anchor. Information is represented using an overarching conceptual framework that seamlessly accommodates society, government, contexts, processes, goals, networks, ecosystems, and phenomena. The framework can operate harmoniously with strategic drivers, change drivers, and efficiency drivers and capitalize on pivotal opportunities spanning technology, policy, strategy, and philosophy resulting in sustainable value. We have looked at 2 case studies. One, Managing Bangalore Roads which impacts people on a day-to-day basis. Another, advocating decentralized economic development using the Capacity-Capability-Opportunity-Potential Model that has a long-term horizon. Our conviction is that democracies should align themselves with Dharma (righteousness) and once that happens everything falls into place [41].

References

1. Al-Kibsi, G., de Boer, K., Mourshed, M., & Rea, N. P. (2001). Putting citizens on line, not in-line. *McKinsey Quarterly, 2*, 65–73.
2. Sakala, http://www.sakala.kar.nic.in/Index. Accessed on July 10, 2022.
3. UN E-Government Knowledgebase, https://publicadministration.un.org/egovkb/en-us/About/ Overview/E-Participation-Index. Accessed on July 10, 2022.
4. Fath-Allah, A., Cheikhi, L., Al-Qutaish, R., & Idri, A. (2014). E-Government maturity models: A comparative study. *International Journal of Software Engineering and Applications., 5*, 71–91. https://doi.org/10.5121/ijsea.2014.5306
5. Malodia, S., Dhir, A., Mishra, M., & Bhatti, Z. A. (2021). Future of e-Government: An integrated conceptual framework. *Technological Forecasting and Social Change, 173*, 121102, ISSN: 0040-1625, https://doi.org/10.1016/j.techfore.2021.121102
6. Zheng, T., & Zheng, L. (2013). Examining e-government enterprise architecture research in China: A systematic approach and research agenda. *Government Information Quarterly, 30*(1), S59–S67, ISSN 0740-624X. https://doi.org/10.1016/j.giq.2012.08.005
7. Helali, R., Achour, I., Jilani, L., & Ben Ghezala, H. (2011). A study of e-government architectures. 78, 158–172. https://doi.org/10.1007/978-3-642-20862-1_11
8. Sedek, K. A., Sulaiman, S., & Omar, M. (2011). A systematic literature review of interoperable architecture for e-government portals. In *2011 5th Malaysian Conference in Software Engineering, MySEC 2011*, pp. 82–87. https://doi.org/10.1109/MySEC.2011.6140648
9. Al-Khanjari, Z., Al-Hosni, N., & Kraiem, N. (2014). Developing a service oriented e-government architecture towards achieving e-government interoperability. *International Journal of Software Engineering and Its Applications (IJSEA), 8*. https://doi.org/10.14257/ ijseia.2014.8.5.04
10. Cellary, W. & Strykowski, S. (2009). e-Government based on cloud computing and service-oriented architecture. In *ICEGOV '09: Proceedings of the 3rd International Conference on Theory and Practice of Electronic Governance*.
11. Rombach, D., & Steffens, P. (2009). e-Government. In S. Nof, (eds.), *Springer Handbook of Automation. Springer Handbooks.* Springer. https://doi.org/10.1007/978-3-540-78831-7_92
12. Basu, S. (2004). E-government and developing countries: An overview. *International Review Of Law Computers & Technology, 18*, 109–132. https://doi.org/10.1080/13600860410001674779
13. Crishantha, N. (2015). Enterprise architecture in the current e-Government context in Sri Lanka, August 2015, *Enterprise architecture in the current e-Government context in Sri La... (slideshare.net)*.
14. Mose, S. M., & Kimani, S., e-Government architecture model for government-to-government deployment of interoperable systems, (A case study of county and national government in Kenya). *International Journal of Innovative Research in Technology & Science(IJIRTS)*, pp. 27–34, ISSN:2321–1156.
15. e-Government Agency, Tanzania. e-Government Application Architecture—Standards and Technical Guidelines, Document Number eGA/EXT/APA/001, November 2017.
16. Kumar, M., Bhatt, R., Vaisla, K. (2015). *Central Architecture Framework for e-Governance System in India Using ICT Infrastructure*, pp. 708–713. https://doi.org/10.1109/ICACCE.201 5.53
17. Ministry of Electronics and Information Technology, Government of India, IndEA Framework (India Enterprise Architecture Framework) October 2018. *India Enterpise Architecture (IndEA) | Ministry of Electronics and Information Technology, Government of India (meity.gov.in).* Accessed on July 10, 2022.
18. Government of Andhra Pradesh, E-Pragati, https://e-pragati.in. Accessed on July 10, 2022.
19. Satyanaryana, J. (2004, January 1). *e-Government: The Science of Possible*. Prentice Hall India Learning Private Limited. ISBN-10: 9788120326088.
20. Shanti Parva, https://en.wikipedia.org/wiki/Shanti_Parva
21. Mahabharat Pancham Khand (Shanti Parva), https://archive.org/details/in.ernet.dli.2015. 342300

22. Rangarajan, L. N. (1992). *Kautilya-The Arthashastra*. Penguin Books India, ASIN: B000OJ8E4G.
23. Bosworth, A. B. (1996). The historical setting of Megasthenes' Indica. *Classical Philology, 91*(2), 113–127.
24. Durant, W. (2017). The Case for India, Gyan Publishing House. ISBN-10812129035X.
25. Panel of Experts, Kautilya's Arthashastra, The way of Financial Management and Economic Governance, Bottom of the Hill Publishing (1 June 2010), ISBN-10: 1935785249.
26. Gautam, P. K. (2019). The Nitisara of Kamandaka, Continuity and Change from Kautilya's Arthashastra, IDSA Monograph Series, No.63.
27. Government of India. (2009). Second Administrative Reforms Commission. Citizen Centric Administration, The Heart of Governance. Twelfth Report.
28. Prabhu, S. M. (2022). Good Governance Model for India. https://www.researchgate.net/public ation/325544499_Good_Governance_Model. Accessed on July 10, 2022.
29. Acemoglu, D., Robinson, J. (2012). *Why Nations Fail: The Origins of Power, Prosperity and Poverty*. Crown.
30. McLeod, S. A. (2022, April 04). *Maslow's Hierarchy of Needs*. Simply Psychology. www.sim plypsychology.org/maslow.html
31. Irwin, M. (2017, July 24). Mastering the pivot: shifting strategy toward opportunity, bottle rocket blog. http://www.bottlerocketadvisors.com/bottle-rocket-blog/2017/7/24/master ing-the-pivot-shifting-strategy-toward-opportunity. Accessed on July 10, 2022.
32. Zachman, J. A. (2003). *Zachman Framework, A Primer for Enterprise Engineering and Manufacturing*. Available online at http://www.zachmaninternational.com
33. Santos, P. S., Almeida, J. P. A., & Guizzardi ,G. (2013). An ontology based analysis and semantics for organizational structure modeling in the ARIS method. *Information Systems Journal*.
34. Codd, E. F. (1970). Relational model of data for large shared data banks. *Communications of ACM, 13*(6).
35. Kaplan, R. (2010). *Conceptual Foundations of the Balanced Scorecard*, Working Paper.
36. Bartels. (1968, January). The general theory of marketing. *The Journal of Marketing, JSTOR*.
37. Weiss, C. (1995). Nothing as practical as good theory: Exploring theory-based evaluation for comprehensive community initiatives for children and families. In Connel, K. & Scorr, (eds.), *"New' Approaches To Evaluating Community Initiatives, Concepts, Methods, and Contexts*. The Aspen Institute, ISBN 0-89843-167-O.
38. TM Forum Frameworx, https://www.tmforum.org/. Accessed on July 12, 2022.
39. Prabhu, S. M., & Subramanyam, N. (2021). Transforming India's social sector using ontology-based Tantra framework. In M. S. Kaiser, J., Xie, & V. S. Rathore (eds.), *Information and Communication Technology for Competitive Strategies (ICTCS 2020)*. Lecture Notes in Networks and Systems, vol. 190. Springer. https://doi.org/10.1007/978-981-16-0882-7_69
40. Prabhu, S. M., & Subramanyam, N., *Analysis of Indian Agricultural Ecosystem using Knowledge-based Tantra Framework*, arXiv:2110.09297. Accessed on 2021, October 13.
41. Prabhu, S. M. (2017). *Democracy and Dharma, LinkedIn Pulse*. https://www.linkedin.com/ pulse/democracy-dharma-shreekanth-m-prabhu/

Chapter 4
Struggling Against Tax Fraud, a Holistic Approach Using Artificial Intelligence

Christophe Gaie

Abstract Nowadays, there are multiple actions to fight against tax frauds. While some articles aim to optimize fraud detection, others focus on recovery but there is no overall approach, to the author's knowledge. This partly explains why fraudsters still flourish and government actions prove ineffective. This chapter presents a comprehensive approach to struggle against tax fraud and optimizing e-Government. The proposal consists in introducing a holistic four-step framework composed that ensures continuous improvement. First, data held by the Tax Administration is reorganized and linked together to enable detection of cross-domain frauds. Then, a model is proposed to improve fraud detection by taking advantage of the previous organization of data. Subsequently, a new stage of resource optimization is proposed to take into account the skills and workload of tax auditors. Finally, a method for comparing and analyzing the results of fraud detection is detailed to ensure a continuous improvement of the proposed system. The proposed model does not aim to maximize the probability to detect fraud but rather to optimize resource allocation and improve the total efficiency and recovery of the anti-fraud system. This new approach is better suited to the concrete organization of government services because human resources are limited and tax verification is time-consuming. This highlights the value of proposing a holistic process to enhance e-Government in Tax Administrations. The current proposal establishes guidelines for a long-term research cycle that will implement and compare different algorithms and configurations. The results obtained will provide details on the comparative performance of the selected implementations.

Keywords Fraud detection · e-Government · Data science · Data analytics · Artificial intelligence · Optimization · Neural networks

C. Gaie (✉)
Centre interministériel de services informatiques relatifs aux ressources humaines (CISIRH), 41 Boulevard Vincent Auriol, 75013 Paris 13, France
e-mail: christophe.gaie@gmail.com

© The Author(s), under exclusive license to Springer Nature Switzerland AG 2023
C. Gaie and M. Mehta (eds.), *Recent Advances in Data and Algorithms for e-Government*,
Artificial Intelligence-Enhanced Software and Systems Engineering 5,
https://doi.org/10.1007/978-3-031-22408-9_4

4.1 Introduction

The fight against fraud is a major concern of governments since it contributes to improving the recovery of state revenue in a context of high public deficits and debts. Concretely, the percentage of the ratio of Debt to the Gross Domestic Product (GDP) of OECD governments reaches 90% in 2020 [1]. In addition, the amount of tax fraud reaches peak levels; for example the European Commission estimated tax evasion at 1 trillion euros in a recent publication [2].

Thus, there is an urgent need to improve the methods for preventing, detecting, measuring, and sanctioning frauds. These fields of interest involve multiple actors with diverse concerns and goals such as citizens, companies, auditors, managers, institutional actors (banks, lawyers) … but also fraudsters themselves! This explains the author's proposal to establish a **HOLISTIC approach**.

First of all, the author points out that optimizing e-Government tax recovery requires distinguishing fraud from the tax gap which is a broader approach as detailed by the International Monetary Fund [3]. The tax gap combines both the compliance gap (i.e. fraud and involuntary non-compliance) and the policy gap (i.e. reductions, exemptions, thresholds, etc.). The holistic approach of this chapter contributes to a better understanding of the limits to be respected and improve stakeholders' ethics by introducing an evasion gap (i.e. legal tax avoidance). Another objective is to offer perspectives to clarify tax regulation (Fig. 4.1).

There are indeed a large number of contributions that tackle a specific area of fraud, especially tax fraud detection, but there is little research that addresses the problem in its entirety. In the present chapter, a comprehensive method is detailed to structure how Tax Fraud should be optimized by government administrations. This proposal leans on previous research concerning Value Added Tax (VAT) securitization with blockchain [4], detecting fraud through Artificial Intelligence (AI) [5], and improving data exchange between public services [6].

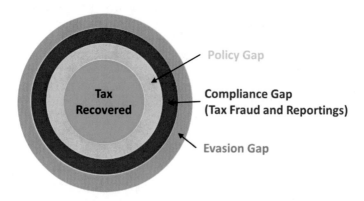

Fig. 4.1 Illustration of tax gap composition

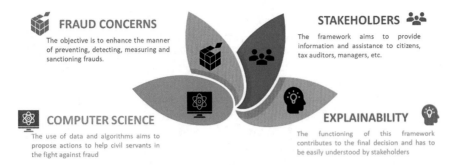

Fig. 4.2 Different dimensions of the holistic framework proposed

The current chapter details how to combine them to enhance the fight against tax fraud that is completely transposable for a different governmental policy. Combining the multiple dimensions detailed previously may result in proposing complex decisions that should be explained to every stakeholder. Thus, the overall approach proposed in this chapter cares to explain the proposals made by the system and to leave the final decision to real people, that is to say, civil servants endowed with human sensitivity and a pragmatic experience. The framework proposed in this chapter is illustrated in the Fig. 4.2.

The different dimensions aim to combine and offer a global perspective in the fight against fraud. To the author's knowledge, this approach is original compared to the existing literature where a focus is made either on the objective or the fraud detection method. The core idea is to benefit from these different views to optimize fraud detection as:

- **Computer science** offers a wide scope of tools to combine data and optimize the research of fraud through data mining, machine learning, or even newer methods.
- **Fraud concerns** underline the importance of preventing, detecting, and sanctioning frauds beyond the simple task of fraud detection.
- **Stakeholders** are fully engaged in the framework proposed as science is dedicated to facilitating their work but not replacing them (like an X-ray helps the medical practitioner).
- **Explainability** guides the construction of the framework because the research and decision criteria must be explained to justify the tax auditor's decision to set a penalty and its amount.

4.2   Literature Review

The literature review first details the different fraud concerns. Next, it describes various fraud detection techniques. Finally, a focus is made on tax fraud which is precisely the subject of this chapter.

Thus, organizing government services requires to define a balance between the effort toward multiple concerns such as preventing tax fraud or its detection and sanction. The digitalization of money exchanges has offered the opportunity to introduce blockchain in the tax system to improve its reliability [7]. The research underlines that the VAT system is well suited to the introduction of blockchain which could increase transparency, compliance, and transaction costs. While the paper indicates the possibility to combine blockchain and cryptocurrencies, this direction seems premature in the context of e-Government procedures.

The combination of data regarding fraud characteristics and transactions improves the results of fraud detection systems [8]. CoDetect is a fraud detection framework that makes it easier to fight against financial fraud. The experimentations carried out offer valuable results thanks to the use of mock data and actual data. Thus, combining different approaches and data improves fraud detection, especially for money laundering in the studied research.

Then, an innovative fraud detection technique consists in detecting potential fraudsters by looking for their hidden links with suspicious taxpayers or tax havens [9]. The combination of Social Network Analysis and machine learning offers valuable results to detect fraud risks and identify the population to control. Interestingly, the analysis helps in the detection of underlying patterns and contributes to the fight against money laundering and terrorism. The usage of the real Spanish database demonstrates the relevance of risk indicators such as the control distance or the number of companies interposed between beneficiaries and wealth.

An interesting approach in detecting tax fraud consists in identifying patterns and subsequently distinguishing taxpayers and potential fraudsters. This approach may be achieved by combining K-mean clustering algorithms with decision trees in order to train a Multilevel Neural Network [10]. Whereas the method is promising the authors do not provide results that could be compared to different tax detection techniques.

A very large number of techniques have been developed to detect fraud such as graph-based anomaly detection (GBAD) approaches [11]. These methods aim to identify potential anomalies in graph-based data, and then detect fraud. The proposed classification framework facilitates the understanding of detection techniques as well as their applicability in different fields. They can be applied and adjusted to enhance e-Government procedures.

A valuable method to detect tax fraud consists in replacing traditional supervised machine learning methods to reduce the effort of labeling data. To this end, de Roux et al. proposed to consider Unsupervised Machine Learning to detect under-reporting of tax declarations [12]. They proposed a model based on both clustering of similar tax declarations and scattering the distribution of declared income. This improves the detection of under-reporting taxpayers for the Urban Delineation tax in Bogotá, Colombia. While the results are encouraging, there are not compared to labeled data and require further study.

Over the last decade, many methods were also proposed to detect financial fraud through data mining [13]. The most popular technique is the support-vector machine (SVM) before Naïve Bayes and Random Forest. The paper also points out that the performance of supervised learning approaches is better than unsupervised. The main reason comes from the difficulty of defining outliers without human intervention. The paper provides a comprehensive classification of data mining techniques against financial fraud but does not address the issue of tax fraud.

Another literature review was carried out through the prism of the Fraud Triangle Theory [14], that is, the combination of opportunity, incentive, and rationalization. The publication examines recent fraud detection approaches that incorporate the fraud triangle in addition to machine learning and deep learning techniques. The authors proved that the combination of approaches is not widely addressed. This confirms our ambition to tackle tax fraud more broadly.

The fight against fraud is a permanent race between fraudsters and tax auditors. As the traditional approaches rely either on the fraud triangle or fraud report data, a new type of fraud based on related-party transactions (RPTs) has recently flourished. Thus, researchers propose new methods to detect abnormal RPTs that frequently appear in financial fraud cases [15]. The results are promising and the RPT graphs improve financial fraud detection. The driver factors identified are loan-based RPTs, high transaction amounts, and the total number of RPTs.

A very innovative tool based on Continuous Wavelet Transform was imagined to estimate the relationship in time-frequency between fraud indicators and tax collection [16]. This approach makes it easier to understand trends concerning tax compliance. As a matter of fact, the general trend is that increasing taxes tends to decrease tax collection. However, high economic growth may offset fraudulent behaviors. This paper also highlights the importance to combine technology and data analysis through human experience to optimize the fight against fraud.

Finally, the focus on tax fraud requires paying attention to VAT fraud detection techniques [17]. Identifying a potential fraudster requires considering several concerns such as the difference between the reference database and the complete population, the specificities of various sectors of activities, the scalability of proposed algorithms, and the evaluation of the performance. The use of unsupervised anomaly detections offers valuable performance, as verified by the proposed lift and hit rates indicators. The use of sectorial information improves fraud detection by taking advantage of varying market conditions and tax law rules in Belgium. While this research is very interesting, it should be further extended to other types of frauds (personal income, real estate, capital ownership, etc.).

Another interesting approach consisted in introducing neural networks to detect tax fraud on personal income tax returns in Spain [18]. The authors adapted the MultiLayer Perceptron (MLP) described in [19] to detect fraud on personal income. The method efficiency is related to the linear structure of every layer. The proposal offers a useful modeling of the likelihood of fraud in Spain and reaches a high detection rate of 84.3%. However, the framework cannot be reused easily in a holistic context as data and algorithms are dedicated to fraud on personal income and are not open source.

A combination of dimension reduction and data mining techniques has recently been proposed to detect fraud on personal income in Spain [20]. The method relies on a database developed by the government services and composed of 2,161,647 records of the 2013 financial fiscal year. Each tax record corresponds to a form with 100 pieces of information which is reduced into 48 principal components representing 80% of the total variance of the group. Then, the use of the Multilayer Perceptron (MLP) model whose results were injected into a decision tree algorithm. The results consist of a segmentation of taxpayers according to their percentage of potential fraud. The analysis of the results obtained shows that the most relevant tax detection variables are the expenditure of Tax Credits, the Tax Credits for investments as well as the rent deductions. This information contributes to the explain ability of the algorithms and offers the opportunity to improve the fiscal law.

The increasing complexity of the fiscal law and the objective to fight against tax evasion have contributed to the tremendous growth of data held by Tax Administrations. Thus, design methods that select features to detect fraud while preserving efficiency are emerging [21]. In this research, the proposed framework proved that reducing the fraud detection dataset from 31 to 5 features offers better efficiency. This conclusion reveals very significant and explains the interest of focusing on the major fiscal information detained, especially that validated by tax auditors.

Improving fraud detection mechanisms requires taking advantage of existing techniques as well as devising new ones. In [22], the authors describe three methods to detect fraud using data mining techniques (outliers in a distribution, anomalies in the search context, and inconstancy from different perspectives). It is also really important to introduce new detection approaches such as large-scale data analysis or social media mining. These methods require a strong evolution of managerial skills to ensure their adoption by tax auditors as a working tool with which they will detect more fraudsters.

Taking into account concrete use cases of tax fraud enables refining the existing classification [23]. It also confronts researchers with real situations with a limited number of observations and heterogeneous data. In the cited paper, the authors propose a framework based on a gradient-boosting algorithm that highlights that rule-based approaches can improve a large scope of fraud detection models. Additionally, the authors prove that using the network of a company increases the ability to detect potential fraud for new companies.

A very innovative approach to fight against tax fraud consists in using machine learning to prevent tax fraud [24]. Using the MultiLevel Perceptron (MLP) neural network allows to estimate the probability that taxpayers will comply with the fiscal law. Interestingly, the research highlights that optimizing tax compliance requires paying attention to patterns that promote ethical behavior. The use of MLP could be further improved by testing it with various numbers of hidden layers or choosing unsupervised neural model.

To conclude this section, the following tables summarizes the contributions and limits of the papers in the context of the proposed holistic approach (Table 4.1):

Table 4.1 Contributions and limits of the different paper cited in the literature review

Paper	Author	Contribution
[4]	Gaie	Prevents fraud by introducing a VAT blockchain
[5]	Gaie	Introduces a mechanism to detect fraud on personal income
[7]	Dourado	Struggles against using a VAT blockchain
[8]	Huang	Detects fraud using mining techniques
[9]	González	Combines social network analysis and machine learning
[10]	Shukla	Combines K-mean clustering algorithms with decision trees to train a Multilevel Neural Network
[11]	Pourhabibi	Provides a review of graph-based anomaly detection (GBAD) approaches
[12]	deRoux	Proposes an Unsupervised Machine Learning to detect under-reporting of tax declarations
[13]	Al-Hashedi	Provides a review of methods to detect financial fraud through data mining
[14]	Sánchez-Aguayo	Another literature review was performed through the prism of the Fraud Triangle (opportunity, incentive, and rationalization)
[15]	Mao	Proposes new methods to detect abnormal related-party transactions that frequently appear in financial frauds
[16]	Monge	Proposes a Continuous Wavelet Transform to estimate the relationship in time–frequency between fraud indicators and tax collection
[17]	Vanhoeyveld	Provides VAT fraud detection techniques especially with outliers
[18]	PérezLópez	Employs neural networks to detect tax fraud concerning personal income
[20]	Vasco	Proposes a method that combines dimension reduction and data mining techniques to detect fraud concerning the Spanish personal income tax
[21]	Matos	Outlines methods that select features to detect fraud while preserving computation efficiency
[22]	Bao	Details three methods to detect fraud using data mining techniques. It also tackles the adoption by tax auditors

4.3   Holistic Process to Improve e-Government Procedures Against Tax Fraud

In this section, a new holistic process is proposed to improve e-Government procedures to fight against tax fraud. This aims to reduce the tax gap (Fig. 4.1) and to fulfill the multiple dimensions considered in the holistic framework (Fig. 4.2) by introducing a continuous improvement method based on four successive stages:

(A) **Organize and Link Data**: the existing approaches used to fight against tax fraud usually aim to detect a specific fraudulent behavior (under-reporting of personal income, non-reporting of VAT, under-estimation of heritage, etc.). This approach is inspired by the human audits that existed before the emergence of e-Government. While this approach is interestingly based on auditors' experience, it does not necessarily take advantage of the cross-referencing of data made possible by digitalization.

(B) **Improve Fraud Detection Mechanism**: the novel structuration of data offers the possibility to optimize the fraud detection algorithms by taking advantage of new dimensions of analysis. At the same time, new algorithms have emerged to identify potential fraud. These algorithms offer better performance than classic neural network techniques. The holistic process outlines that the detection algorithms must be continuously adjusted to benefit from the new structure of data.

(C) **Optimize Resource Allocation**: this step aims to optimize the allocation of tax auditors to maximize fraud detection and recovery while minimizing the resource dedicated to this objective. This task is rarely addressed in the literature because the research generally focuses on optimizing the detection of frauds. However, the improvement of e-Government services requires providing a tax schedule adjusted to the availability and skills of auditors.

(D) **Analyze Results**: the fight against fraud is a constant game of cops and robbers. Each of them modifies their actions either to increase the amount of money diverted or the amount of fraud detected and recovered. Therefore, the results obtained contribute to the continuous optimization of the fraud detection process (Fig. 4.3).

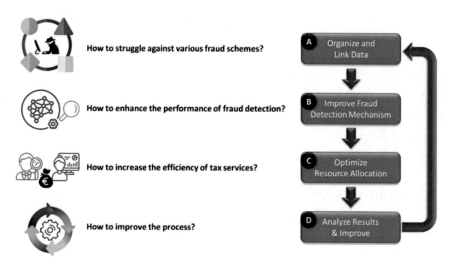

Fig. 4.3 Sequential steps to implement the proposed framework

4.3.1 Organize and Link Data

The classic approach toward fraud detection usually relies on separate applications that aim to manage specific taxes (income, heritage or VAT). Historically, fraud detection has been built on top of the existing applications as an additional process designed by tax experts. Thus, the original data structuration makes it possible to tackle specific tax fraud schemes, that is to say, fraud patterns that are already known by the Tax Administration.

The new decompartmentalized data organization and linkage aims to enable a cross-domain approach as well as to take advantage of data science to detect new fraud schemes. The figure below illustrates how this approach offers new opportunities to enrich fraud detection (Fig. 4.4).

The paragraphs below detail a model that corresponds to the data structuration and linkage principles for people (companies are deliberately excluded from the model). First, let's start by defining the following variables:

- $r_{i,c,y}$ corresponds to the revenues of a person i obtained from a company c and declared for the year y
- $b_{i,y}$ corresponds to the total value of bank accounts of a person i declared for the year y
- $v_{i,y}$ corresponds to the cumulative value of the properties of a person i declared for the year y
- $s_{i,c,y}$ corresponds to the value of shares that a person i holds in the company c declares for the year y

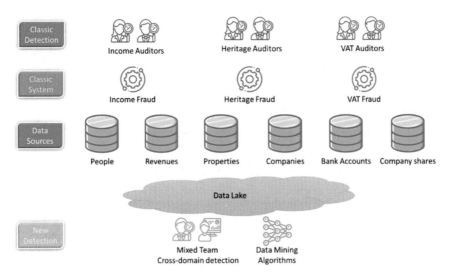

Fig. 4.4 A novel decompartmentalized data structuration and linkage

- $p_j(t_i)$ corresponds to the amount of tax recovered when a person i is audited by the tax verificator j with the tax process p() and the tax situation[1] t_i.

Then, let's introduce a very simplified model that relies on the three following dimensions:

- Cumulative revenues and bank accounts $\sum_{c,y} r_{i,c,y} + \sum_y b_{i,y}$
- Cumulative property value $\sum_y v_{i,y}$
- Cumulative share amount $\sum_{c,y} s_{i,c,y}$.

The tax situation can be expressed as follows:

$$t_i = \left\{ \alpha_i \sum_{c,y} r_{i,c,y}; \ \beta_i \sum_y b_{i,y}; \ \gamma_i \sum_y v_{i,y}; \ \delta_i \sum_{c,y} s_{i,c,y} \right\} \tag{1}$$

Thus, a typical objective is to identify the potential fraudster i to assign to tax auditors j in order to maximize the amount of tax recovered for a given tax process can be expressed as:

$$\arg\max_{\{i,j\}} \sum_j p_j \left(\left\{ \alpha_i \sum_{c,y} r_{i,c,y}; \ \beta_i \sum_y b_{i,y}; \ \gamma_i \sum_y v_{i,y}; \ \delta_i \sum_{c,y} s_{i,c,y} \right\} \right) \tag{2}$$

The parameters $\alpha_i, \beta_i, \gamma_i, \delta_i$ correspond to the weight of dimensions that tax auditors must consider in phase C of the process. Using the modelization proposed in Fig. 4.3, the four steps of the process may be illustrated as follows (Fig. 4.5):[2]

Finding the optimal combination of people {i} and tax auditors {j} to recover the highest amount of fraud is a classic optimization process. Paragraph B describes how to detect the potential fraudsters {i} while paragraph C characterizes how to assign auditors {j} efficiently. The method to solve this problem relies on convex optimization and is detailed in [25].

4.3.2   Improve Fraud Detection Mechanism

A decisive concern for detecting fraud is to identify unconventional situations. Anyone can easily understand that raising a lot of money is easier when taxpayers earn a lot of money rather than when they are unemployed. At the same time, the detection of unconventional tax status does not necessarily correspond to fraud. For instance, owning a luxurious estate may be explained by a classic inheritance.

Another dimension to consider is the tendency of fraudsters to conceal their fraud, this can be solved by random verifications to ensure a permanent threat of illegal

---

[1] An audit is usually performed in the last three years.
[2] $\alpha_i, \beta_{i,}, \gamma_i, \delta_i$ considered equal to 1.

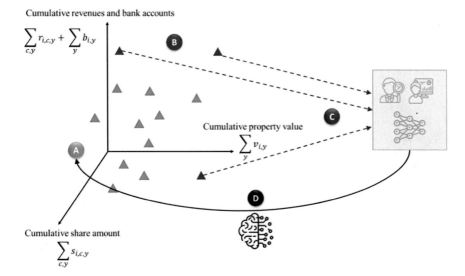

Fig. 4.5 Illustration of the holistic fraud detection approach

behaviors. This method of detection has always been employed by the police to arrest robbers or by customs officers against drug dealers.

Finally, there should be a systematic verification of tax situations presenting major financial issues. Taxpayers with huge financial means tend to optimize their tax return with a risk of excessive tax optimization or misinterpretation of the law. Beyond the mathematical considerations, this verification may also ensure the respect for ethics where the wealthiest people comply with their fiscal obligations. The figure below illustrates the different populations to consider when researching fraud (Fig. 4.6).

This representation of groups or clusters that enable to identify outliers has already been used in the context of tax fraud detection [26] and [27]. This approach may be adjusted to the detection of fraud in a n-dimension context as illustrated in this chapter.

It is useful to note that optimizing the probability to detect a fraud requires finding the combination of parameters $(\alpha_i, \beta_i, \gamma_i, \delta_i)$ which significantly spread the fiscal files and outline the atypical tax situations. This job is usually performed by a mixed team including senior tax auditors and data scientists.

4.3.3   Optimize Resource Allocation

In the previous step, the objective was to optimize the probability of detecting a person with a high probability of fraud as well as a high amount to potentially recover, considering that auditors had the same skills. However, the concrete situation within

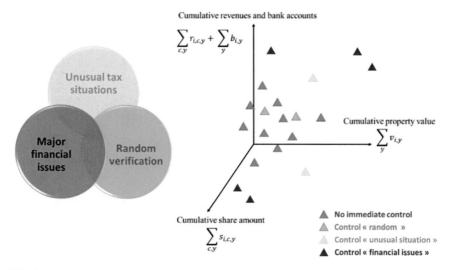

Fig. 4.6 Optimizing tax audits to perform

government services requires taking advantage of the heterogeneity of skills held by tax auditors.

In addition, the number of files that a tax auditor can simultaneously examine is limited to a given parameter M_j and its efficiency may be improved according to this ability skill. The following figure illustrates the diversity that the amount of tax recovered when a person i is audited by the entitled tax verificator j is d $p_j(t_i)$ and depends on the auditor's skills (Fig. 4.7).

Thus, the problem can now be expressed as follows:

- $i = 1 \dots N$, corresponding to the previously selected files
- $S_j(d)$ is the skill of the tax auditor to deal with the four dimensions of tax fraud d
- $e_i(j)$ is the effort made by the tax auditor j to control the tax situation i
- E_j is the maximum effort that the tax auditor j can provide.

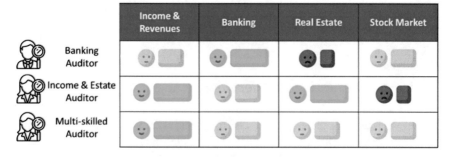

Fig. 4.7 Illustration of the different skills detained by auditors

Consequently, the model may be refined such that:
- $p_j(t_i, \{S_j(d)\})$ corresponds to the amount of tax recovered when a person i is audited by the tax verificator j with the skills $\{S_j(d)\}$ using the tax process $p()$ on the tax situation t_i.

Finally, the problem to solve is:

$$\arg\max_{\{i,j\}} \sum_j p_j(t_i, \{S_j(d)\})$$

$$\text{where for each tax auditor } j, \sum_i e_i(j) \le E_j \qquad (3)$$

Note that a concrete implementation can be obtained by using the following formula:

$$p_j(t_i, \{S_j(d)\}) = p_j\left(S_j(r) \cdot \alpha_i \sum_{c,y} r_{i,c,y} + S_j(b) \cdot \beta_i \sum_y b_{i,y}\right.$$

$$\left. +S_j(v) \cdot \gamma_i \sum_y v_{i,y} + S_j(s) \cdot \delta_i \sum_{c,y} s_{i,c,y}\right) \qquad (4)$$

4.3.4 Analyze Results

The holistic approach proposed is based on a continuous improvement process. This requires building a learning database to perform fraud detection, optimize resources, analyze results and optimize algorithms. As resources are limited, it is likely that the optimization will not be performed on identical cases but rather on similar cases.

Thus, the results obtained using the classical confusion matrix [28] should be analyzed by experts in a mixed team composed of experts and data scientists. Indeed, after a first phase where parameters can evolve significantly, the algorithms should be stabilized and should only evolve with small steps (Fig. 4.8).

The optimization of the implemented algorithms must take into account the fraud detection rate given by the confusion matrix but also the effort employed to obtain the results as well as the effective money recovered (generally low as indicated in [29]). The holistic approach aims to enhance every e-Government dimension to fight against tax fraud. Thus, it aims to improve tax compliance, recovery of evaded taxes as well as government efficiency.

A trade-off can be mathematically modeled by refining the expression (2.22) defined in the previous section but the experience of senior tax auditors should be prefered in the first implementation. Indeed, it offers the opportunity to highlight the importance of human expertise and offer better explain ability in the options chosen.

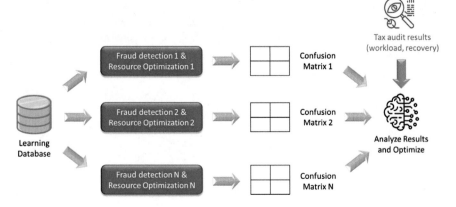

Fig. 4.8 Method to optimize the fraud detection and resource optimization system

4.4 Conclusion and Perspectives

This chapter has proposed a holistic approach to optimize the fight against tax fraud detection. Indeed, the government services need to detect fraud and simultaneously optimize their human resources as well as increase tax recovery. These last two dimensions are often relegated to the background in the fight against tax fraud. Accordingly, the author has devised a four-step approach that addresses all dimensions of the fight against tax fraud: (1) it organizes and models data, (2) it mathematically describes how to detect tax fraud, (3) it optimizes available audit resources and (4) it finally ensures continuous improvement of the system.

This research constitutes a new stage in the approach of the author towards strengthening the e-Government methods to fight against frauds because it combines multiple former methods in a common framework and offers new perspectives. The multiple dimensions are combined in a common scientific approach based on a mathematical model and the experience of tax auditors. This choice aims to build an operational system easy to adopt by civil servants. It aims to help them in their daily work and prevent them from fearing the disappearance of their jobs.

Future research will be dedicated to the implementation of the proposed model and the investigation of the most efficient algorithms to implement. This work will require a significant effort to build a test database, implement the algorithms, calibrate the parameters and analyze the results that will be obtained.

Acknowledgements The author would like to express his gratitude to Bernard Gaie for his valuable suggestions and careful review of the manuscript.

References

1. OCDE. (2022). OECD sovereign borrowing outlook 2022. *Éditions OCDE, Paris.* https://doi.org/10.1787/b2d85ea7-en

2. European Commission. (2017). https://ec.europa.eu/taxation_customs/huge-problem_en, Verified on the 15th of June 2022.

3. The Revenue Administration Gap Analysis Program. *Technical Notes and Manuals, 2021*(009), A001. Retrieved Jun 4, 2022, from https://www.elibrary.imf.org/view/journals/005/2021/009/article-A001-en.xml

4. Gaie, C., & Mueck, M. (2021). A hybrid blockchain proposal to improve value-added tax recovery. *International Journal of Internet Technology and Secured Transactions.* https://doi.org/10.1504/ijitst.2022.119668

5. Gaie, C., & Mueck, M. (2021). Public services data analytics using artificial intelligence solutions derived from telecommunications systems. *International Journal of Business Intelligence and Systems Engineering.* https://doi.org/10.1504/IJBISE.2020.10042244

6. Gaie, C. (2021). An API-intermediation system to facilitate data circulation for public services: The French case study. *International Journal of Computational Systems Engineering.* https://doi.org/10.1504/IJCSYSE.2021.120292

7. Dourado, L., Silva, P., Peres, C., & Díaz, D. (2022). Challenges and opportunities for a fiscal blockchain. *American Academic Scientific Research Journal for Engineering, Technology, and Sciences, 87*(1). https://asrjetsjournal.org/index.php/American_Scientific_Journal/article/view/7520

8. Huang, D., Mu, D., Yang, L., & Cai, X. (2018). *CoDetect: Financial Fraud Detection With Anomaly Feature Detection. IEEE Access, 6,* 19161–19174. https://ieeexplore.ieee.org/stamp/stamp.jsp?arnumber=8325544

9. González, I., & Alfonso, M. C. (2018). Social network analysis tools in the fight against fiscal fraud and money laundering. In *15th International Conference on Modeling Decisions for Artificial Intelligence (MDAI 2018)*, 15–18 Oct 2018, Palma de Mallorca, España, pp. 226–237. ISBN 978-84-09-05005-5. https://oa.upm.es/54733/

10. Shukla, Y., Sidhu, N., Jain, A., Patil, T. B., & Sawant-Patil, S. T. (2018). Big data analytics-based approach to tax evasion detection. *International Journal of Engineering Research in Computer Science and Engineering, 5*(3). https://ijercse.com/abstract.php?id=12267

11. Pourhabibi, T., Ong, K. L., Kam, B. H., & Boo, Y. L. (2020). Fraud detection: A systematic literature review of graph-based anomaly detection approaches. *Decision Support Systems, 133,* 113303, ISSN 0167-9236. https://doi.org/10.1016/j.dss.2020.113303

12. de Roux, D., Perez, B, Moreno, A, Villamil, M. D. P., & Figueroa, F. (2018). Tax fraud detection for under-reporting declarations using an unsupervised machine learning approach. In *Proceedings of the 24th ACM SIGKDD International Conference on Knowledge Discovery & Data Mining (KDD '18).* Association for Computing Machinery, New York, NY, USA, pp. 215–222. https://doi.org/10.1145/3219819.3219878

13. Al-Hashedi, K. G., & Magalingam, P. (2021). Financial fraud detection applying data mining techniques: A comprehensive review from 2009 to 2019. *Computer Science Review, 40,* 100402, ISSN 1574-0137. https://doi.org/10.1016/j.cosrev.2021.100402

14. Sánchez-Aguayo, M., Urquiza-Aguiar, L., & Estrada-Jiménez, J. (2021). Fraud detection using the fraud triangle theory and data mining techniques: A literature review. *Computers, 10*(10), 121. https://doi.org/10.3390/computers10100121

15. Mao, X., Sun, H., Zhu, X., & Li, J. (2022). Financial fraud detection using the related-party transaction knowledge graph. *Procedia Computer Science, 199,* 733–740. ISSN 1877-0509. https://doi.org/10.1016/j.procs.2022.01.091

16. Monge, M., Poza, C., & Borgia, S. (2022). A proposal of a suspicion of tax fraud indicator based on Google trends to foresee Spanish tax revenues. *International Economics, 169,* 1–12, ISSN 2110-7017. https://doi.org/10.1016/j.inteco.2021.11.002

17. Vanhoeyveld, J., Martens, D., & Peeters, B. (2020). Value-added tax fraud detection with scalable anomaly detection techniques. *Applied Soft Computing, 86*, 2020, 105895, ISSN 1568-4946, https://doi.org/10.1016/j.asoc.2019.105895
18. Pérez López, C., Delgado Rodríguez, M. J., & de Lucas Santos, S. (2019). Tax fraud detection through neural networks: An application using a sample of personal income taxpayers. *Future Internet, 11*(4), 86. https://doi.org/10.3390/fi11040086
19. Gardner, M. W., & Dorling, S. R. (1998). Artificial neural networks (the multilayer perceptron)—A review of applications in the atmospheric sciences. *Atmospheric Environment, 32*(14–15), 2627–2636, ISSN 1352-2310, https://doi.org/10.1016/S1352-2310(97)00447-0
20. Vasco, C. G., Rodríguez, M. J. D., & de Lucas Santos, S. (2021). Segmentation of potential fraud taxpayers and characterization in personal income tax using data mining techniques. *Hacienda Pública Española/Review of Public Economics, IEF, 239*(4), 127–157. https://ideas.repec.org/a/hpe/journl/y2021v239i4p127-157.html
21. Matos, T., Macedo, J.A., Lettich, F., Monteiro, J. M., Renso, C., Perego, R., & Nardini, F. M. (2020). Leveraging feature selection to detect potential tax fraudsters. *Expert Systems with Applications, 145*, 113128, ISSN 0957-4174, https://doi.org/10.1016/j.eswa.2019.113128
22. Bao, Y., Hilary, G., & Ke, B. (2020, November 24). Artificial intelligence and fraud detection. In: V. Babich, J. Birge, G. Hilary, (eds.), *Innovative Technology at the interface of Finance and Operations*. Springer Series in Supply Chain Management, forthcoming, Springer Nature, Available at SSRN: https://ssrn.com/abstract=3738618 or https://doi.org/10.2139/ssrn.3738618
23. Baghdasaryan, V., Davtyan, H., Sarikyan, A., & Navasardyan, Z. (2022). Improving tax audit efficiency using machine learning: The role of taxpayer's network data in fraud detection. *Applied Artificial Intelligence, 36*(1), 2012002. https://doi.org/10.1080/08839514.2021.2012002
24. Ioana-Florina, C., & Mare, C. (2021). The utility of neural model in predicting tax avoidance behavior. In I. Czarnowski, R. J. Howlett, & L. C. Jain (eds.), *Intelligent Decision Technologies. Smart Innovation, Systems and Technologies*, vol. 238. Springer. https://doi.org/10.1007/978-981-16-2765-1_6
25. Boyd, S., & Vandenberghe, L. (2004). *Convex Optimization*. Cambridge University Press. https://web.stanford.edu/~boyd/cvxbook/bv_cvxbook.pdf
26. Vanhoeyveld, J., Martens, D., & Peeters, B. (2020). Value-added tax fraud detection with scalable anomaly detection techniques. *Applied Soft Computing, 86*, 105895. ISSN 1568-4946, https://doi.org/10.1016/J.ASOC.2019.105895
27. Savić, M., Atanasijević, J., Jakovetić, D., & Krejić, N. (2022). Tax evasion risk management using a hybrid unsupervised outlier detection method. *Expert Systems with Applications, 193*, 116409, ISSN 0957-4174, https://doi.org/10.1016/j.eswa.2021.116409
28. Stehman, S. V. (1997). Selecting and interpreting measures of thematic classification accuracy. *Remote Sensing of Environment, 62*(1), 77–89, ISSN 0034-4257, https://doi.org/10.1016/S0034-4257(97)00083-7
29. Curti, F., Mihov, A. (2018). Fraud recovery and the quality of country governance. *Journal of Banking & Finance, 87*, 446–461, ISSN 0378-4266, https://doi.org/10.1016/j.jbankfin.2017.11.009

Chapter 5
e-Government and Green IT: The Intersection Point

Rodrigo Franklin Frogeri⊙, **Wendell Fioravante da Silva Diniz**⊙,
Pedro dos Santos Portugal Júnior⊙, **and Fabrício Pelloso Piurcosky**⊙

Abstract Information and Communication Technology (ICT) use in the public sector has changed in recent decades. There is a growing demand for governments to expand their data processing and storage capabilities. Public services provided to citizens are being migrated from physical processes to digital ones. Thus, governments guided by digital technologies have emerged and the concept of e-Government standout. However, technological evolution must development aligned with current environmental concerns. Hereupon, the concept of Green IT has been highlighted in the scientific literature and in organizations/industries that seek the sustainable use of ICT. Thus, this study aims to identify the intersection point between the scientific literature on e-Government and Green IT to enable interdisciplinary research lines involving these subjects. To achieve the objective, first, Systematic Literature Review (SLR) techniques were applied to identify 102 studies. In a second step, techniques of Lexical Analysis and Content Analysis were applied to the textual corpus of the studies identified by the SLR. Third, we applied Zipf's and Bradford's bibliometrics laws to identify the main scientific journals and prolific authors. Our analyses suggest that e-Government and Green IT subjects have no associations in the literature studied. Each of the themes is discussed and analyzed differently, but they share similar theoretical underpinnings in the Information Systems field.

R. F. Frogeri (✉) · P. dos Santos Portugal Júnior
Research Department, Graduate Program in Management and Regional Development, Centro Universitário do Sul de Minas - UNISMG, Varginha, MG, Brazil
e-mail: rodrigo.frogeri@professor.unis.edu.br

P. dos Santos Portugal Júnior
e-mail: pedro.junior@professor.unis.edu.br

W. F. da Silva Diniz
Institute of Technological Sciences, Universidade Federal de Itajubá - UNIFEI, Itabira, MG, Brazil
e-mail: wendelldiniz@unifei.edu.br

F. P. Piurcosky
Entrepreneurship, Research and Extension Center, Centro Universitário Integrado, Campo Mourão, PR, Brazil
e-mail: fabricio.pelloso@grupointegrado.br

© The Author(s), under exclusive license to Springer Nature Switzerland AG 2023
C. Gaie and M. Mehta (eds.), *Recent Advances in Data and Algorithms for e-Government*,
Artificial Intelligence-Enhanced Software and Systems Engineering 5,
https://doi.org/10.1007/978-3-031-22408-9_5

Keywords Green IS · Electronic government · e-Services · IT environmental sustainability · Energy informatics

5.1 Introduction

Information and Communication Technology (ICT) use in the public sector has changed in recent decades [1]. The intersection between the fields of Information Systems (IS)/ICT and public administration is conceptualized as the e-Government [2]. There is a growing demand for governments to expand their data processing and storage capabilities [3]. Public services provided to citizens are being migrated from physical processes to digital ones [4]. Thus, governments guided by digital technologies have emerged and the concept of e-Government standout [5].

However, technological evolution must develop aligned with current environmental concerns [6–8]. Hereupon, the concept of Green IT has been highlighted in the scientific literature and in organizations that seek the sustainable use of the ICT [9–11]. The term "Green IT" was initially applied to differentiate IT artifacts designed with environmental sustainability principles. In a similar vein, the term "Green IS" refers to information systems that can make organizational processes and the products they manage environmentally sustainable [7]. Both terms (Green IT or IS) refer to the use of ICT to promote sustainability [12].

Nevertheless, ICT can have both positive and negative impacts on the environment. Positive impacts may be associated with dematerialization and online delivery of products or services, reduction in travel needs, and greater energy efficiency in the production, use, and recycling of materials [13]. Otherwise, the negative impacts associated with ICT may be associated with the production and distribution of ICT equipment, the energy consumption of equipment (direct consumption and for cooling equipment), the short life cycle of electronic equipment, and the e-waste [13]. Furthermore, the estimated global CO_2 emissions of ICTs (excluding radio and television) are equivalent to 1 Gigaton of CO_2 [14] and digitalization increases energy consumption [15]. On one hand, we have the demand to improve organizations and industries [16], and governments [1] in digitalization. On the other hand, we have the demand to promote sustainable use of ICT and reduce energy consumption.

Understanding the intersection between the concepts of Green IT and e-Government may be a way for governments to digitize sustainably. Green IT can enable the reduced power consumption of States in the context of global warming and the high increase in energy products. Green IT can contribute to better knowledge and control of the government IT system as it tends to rationalize it. Furthermore, Green IT can assist government procedures simplification (e.g., automatic tax statements, automatic subsidy computation, and another automatic processes for the citizens).

Thus, this study aims to identify the intersection point between the scientific literature on e-Government and Green IT to enable interdisciplinary research lines involving these subjects. To achieve the objective, first, Systematic Literature Review

(SLR) techniques were applied to identify 102 studies. In a second step, techniques of Lexical Analysis and Content Analysis were applied to the textual corpus of the studies identified by the SLR. Third, we have applied Zipf's and Bradford's bibliometrics laws to identify the main scientific journals and prolific authors.

5.2 A Historical Review of the e-Government Concept

Since the late 1990s, the increased use of ICTs as a platform for communication and service delivery for citizens and businesses has driven reforms in the public sector [17]. We can define e-Government as the electronic delivery of information and services to citizens, businesses, and public administration [18]. The emerging field was born with high expectations to enhance public management cost efficiency and effectiveness and promote increased engagement with society [19].

As an academic subject, interest in e-Government started to grow in the early 2000s, with many researchers, mostly ICT-oriented, taking a more detailed approach to e-Government concepts. Lips [19] points out researchers started to use and adapt concepts, approaches, and perspectives related to ICT use in the private sector to the government sector, often ignoring unique characteristics and particular nuances of governments. One example is the widespread use of a classification system like that used to classify e-commerce-related concepts, such as Government to Consumer (G2C) mirroring Business to Consumer (B2C), and further analogies such as G2B and G2G. Another example shows up in the widely adopted e-Governments sophistication models, such as the one proposed by Layne and Lin [20], in which a four-stage e-Government maturity model governs the evolution of e-Government as it experiences increasing levels of technological and organizational complexity, going through information provision (cataloging), transaction, and vertical and horizontal integration. The model was a first attempt to offer a common ground for creating and developing e-Government initiatives as the subject was in its infancy at the time.

By the middle of the 2000s-decade, e-Government observed a dramatic growth as a research topic, which can be verified by the consolidation of several e-Government-specific annual conferences, scientific journals, and books [2]. This impressive growth in a relatively short time called for a deep analysis of the philosophies, theories, and methods established in the field. Heeks and Bailur [2] bring a deep analysis of the directions and tendencies at that moment. Heeks and Bailur's [2] work consists of an analysis of 84 papers from three sources. It concludes that, despite most works presenting an optimistic (even over-optimistic) view, the overall quality of the research was poor, with some issues inherited from the "parent" topics, Information Systems and Public Administration, fields often pointed as having philosophical, theoretical, methodological, and practical shortcomings.

By the end of the first decade of the 21st Century and the beginning of the next one, more and more countries were moving to adopt e-Government-based services, despite the supporting theories still being premature. Following the guidelines found in Heeks and Bailur [2], some researchers started to propose new methods based on a tentatively more grounded theoretical framework. Specifically, Shareef et al. [21] recognizes that the fundamental essence of e-Gov adoption is citizen-focused, a common characteristic in all implementations, despite each country having its specifics and variations regarding missions and objectives. The conclusion is that the willingness of citizens to adopt e-Gov services is the determinant factor for the success of an implementation.

However, citizens are not the only consumer of e-Gov services. A considerable range of those users comprises business users. Lee et al. [18] offer a study on the willingness of e-Gov services adoption by business users, providing some interesting insights. Most of the research was focused on the supplier (government) side, over-looking the perceptions of the user side—citizens and businesses. Initial efforts made by Carter and Belanger [22] and Welch et al. [23] studied the e-Gov adoption behavior of individual citizens, leading to Shareef et al. [21] conclusions. Private sector busi-nesses are key economic actors and crucial customers of governmental services. They are also an important source of public revenue. Therefore, business users' adoption of e-Gov services may be an important factor in the success or failure of implemen-tation. The study points out that business users have some common motivators for e-Gov adoption with citizens, like saving time and cost, increased accuracy, relia-bility, etc. Still, the nature of the private sector's relationship with the government is essentially different from the citizens. Particularly, citizens have a political role through voting that is not present in business users. Also, the perception of benefits and risks of adopting e-Gov services differs. Business users' decision-making process is more complex, given that it stems from collective action rather than the individual motivation of a citizen. The study concludes that the willingness to adopt e-Gov services depends on the level of trust the businesses have acquired with traditional brick-and-mortar public services. Trust in Internet technology's ability to provide online services alone is insufficient for adopting e-Gov services, though previous experiences in other types of services, like e-commerce, may have some influence.

Another aspect of e-Gov implementation that lacked study was the perceived quality of services. Papadomichelaki and Mentzas [24] propose a model for measuring e-Gov service quality, derived from well-established models and prac-tices for assessing e-service, site, and portal quality. The study introduces a model called e-GovQual built upon 21 quality attributes under four dimensions: Reliability, Efficiency, Citizen Support, and Trust. The model was validated through an empirical study that questioned citizens over the quality attributes and provided verification and confirmation of the scales designed for the model. The study encourages further use and refinement of the model, supplying e-Gov researchers with a reliable tool for assessing the quality of e-Gov services.

As the technologies providing Web Applications moved to Cloud-based solutions, it was natural to think that e-Gov applications would also follow and migrate to this new architecture. Wyld [25] delivers an investigation of the migration of the e-Gov services to the cloud. The study observes that by the time it was published, several governments were already adopting the cloud and becoming, in many instances, the main force driving the development of the cloud computing sector. After scru-tinizing the migration to the cloud in the USA, Europe, and Asia, the author also presented a six-step "Cloud Migration Strategy" to help governments in shifting to cloud computing. The work then proceeds to conclusions after offering a future assessment of the implications of this migration for public agencies and the IT community.

Following the tendency pointed out by Wyld [25], Liang et al. [1] present an empirical study on China's e-Gov cloud assimilation, offering a view of how the process went on the specifics of the Chinese governmental practices. Their work says that as both academia and industry recognize the impact of value and assimilation of ICTs at an organizational and performance level, government agencies must in choosing various e-Gov cloud assimilation dimensions. They are assimilation depth (the scope and diversity of e-Gov cloud usage) and assimilation breadth (the intensity of e-Gov cloud usage for aligning the actions of government affairs). The conclusions are that though cloud adoption was successful in the private sector, the public sector was falling behind. Nonetheless, the public sector was increasingly adopting cloud solutions to address the challenges of the traditional isolated investment model of ICT infrastructure. Liang et al. [1] also has shown that "the process of e-Gov cloud assimilation on public value creation is much more complex. Agencies seeking to improve public value from e-Gov cloud need to recognize the individual effects of various assimilation dimensions and consider these distinct fit strategies of e-Gov cloud assimilation" (p. 11).

As the theoretical frame and actual practices evolve, we are now enabled to investigate the e-Gov phenomenon in a more fine-grained view, municipality-wise rather than federal or state-wise. Ingrams et al. [26] developed a longitudinal study on the largest cities of the 100 "most wired" countries in the world, compiling data from the biennial e-Governance Global Survey. The period covered by the paper ranges from 2003 to 2016. The study used global data instead of American or European-centric data found in previous works, thus providing an international view of e-Gov initiatives. The analysis focused on the e-Gov adoption maturity of implementations under the Layne and Lee [20] four-stage model. The findings corroborate previous works reinforcing the theoretical framework in construction with an emphasis on interdisciplinarity, converging Information Science with public administration and management. Cluster analysis also has shown that the dimensions of e-Gov develop simultaneously. One particularly notable finding is that the democratic level, which influences privacy and security, seems to have significance only in large countries. In small countries, the democratic level was inversely proportional to the e-Gov development stage. It suggests that small countries are better at providing traditional services. Small non-democratic countries have the advantage of quickly deploying online services, perhaps because they tend to concentrate services in centralized agencies.

Following the discussions that e-Government is an interdisciplinary construct, Malodia et al. [27] conceptualize e-Government as a multidimensional construct with customer orientation, channel orientation, and technology orientation.

Thus, the interactions citizen-to-citizen (C2C)—customer orientation dimension—get integrated into government activities and form e-Government value chains

[28]. Furthermore, e-Governance performance factors that potentially serve as evaluation criteria tend to be overly sensitive to context (e.g., policy area, systemic constellations, institutional settings, and administrative traditions) [29].

5.3   Green IT

In the past decades, we perceived governments and institutions increasingly adopting environmental sustainability practices derived from preoccupations about social responsibility, inequalities of wealth distribution, and natural resources depletions.

Chen et al. [12] introduce one of the first attempts at assessing the roles that IT can assume in pursuing ecological sustainability. To this purpose, the work uses the Institutional Theory as the theoretical framework to build upon leveraging IS resources to achieve the three milestones of environmental sustainability: eco-efficiency, eco-equity, and eco-effectiveness. Finally, it concludes that IS occupies a central position in mitigating institutional pressures by using its traditional roles: automate, inform, and transform, aiming to achieve environmental sustainability.

Watson et al. [30] introduces a new subfield in IS research: energy informatics. They postulate that IS is essential to reduce energy consumption and, therefore, diminish CO_2 emissions. The increased energy consumption derived from fossil fuels causes growth in CO_2 emissions. One can argue that, as IS increases the efficiency of business enterprises, it consequently increases the energy demand, thus assembling a dangerous feedback cycle. However, IS has the potential to decrease energy demands through the application of intelligent design and leveraging chokepoints, thus creating a more efficient process. These are the practices named Green IS.

Although the intention to achieve environmental sustainability is clear, how to proceed is not. Elliot [31] investigates perspectives and proposes a framework for IT-enabled business transformation toward achieving sustainability. The work focuses on business and technology, and research justified by its power to drive innovation and diffusion of technologies worldwide. Business transformation is predicted and required, as they have consistently reshaped whenever some new major force emerges. It happened in the last decades with the globalization and information technology revolutions. Now, climate change surfaces as a new power driving business transformation. The framework development was grounded firmly in General Systems Theory.

As the concerns with environmental issues continued to grow, many governments started to pass a growing number of regulatory devices that put business enterprises under increasing social, legal, and economic pressures to adopt sustainable practices [7]. The new field of Green IT emerged as the response to enable the firms to achieve environmental compliance and manage related institutional risks. Green IS (Green IT-based Information Systems) can direct business processes and products environmentally sustainable [7, 12].

Butler [7] proposes a theoretical model that employs the Institutional Theory to explain how the exogenous components (regulations, norms, and cultural-cognitive factors) shape IT decisions on the design and production of sustainable products. The authors apply the Organizational Theory to describe endogenous practices on Green IS supporting sense-making, decision-making, and knowledge creation centered on environmental sustainability. The paper evidenced the effect of the practitioners on Green IS initiatives.

Dao et al. [8] advocate for a more expansive role for IT, going beyond energy consumption reduction. Their work proposes an integrated sustainability framework and theoretical insights on integrating the different types of IT resources (automate, information, transform, infrastructure). The study concludes that developing said capabilities will serve the environment and people and generate value that could enhance profitability and sustained competitive capacity.

Bose and Luo [32] introduce a theoretical perspective on potential Green IT initiatives via virtualization. They noticed a gap in the literature that lacked a theoretical framework for implementing Green IT initiatives using modern technologies such as virtualization. Additionally, the work argues that virtualization is the primary force for organizations to integrate environmental sustainability into business and IT practices and hence advocates that virtualization is a viable means through which an organization can implement practices centered on the global green mantra "Reduce, Reuse, and Recycle" [32].

Landum et al. [33] suggest a framework for the alignment of ICT with Green IT based on nine phases. Phase 0 refers to the problem identification; phase 2 refers to the evaluation and continuous improvement; phase 3 is about study and planning; phase 4, telecommunications and printing; phase 5—information Security; phase 6—innovation; phase 7—improvement of citizen services; and phase 8—evaluation/opinion. Meanwhile, Verdecchia et al. [34] discuss the horizons (e.g., today, next four to six years, and beyond six years) and types of Green IT solutions. Four types of solutions have been identified by the authors namely: technical, social, and environmental solutions, and paradigm shifts. Solutions toward Green IT involve moving to the cloud on a current horizon and designing for reuse, hardware breakthroughs, and high-density storage beyond six years' horizon. Anthony et al. [35], investigated the factors that influence IT professionals' and IT managers' intention to deploy Green IT/IS practices. According to the authors, the environmental-friendly initiatives derived from Process-Virtualization Theory (PVT) and Perceived Organizational e-Readiness Theory (POER) help IT practitioners and IT managers understand the potential benefits associated with Green IT/IS practices deployment in Collaborative Enterprise (CE).

5.4 Methodology

Methodologically, the study has a qualitative and interdisciplinary approach, inductive logic, and interpretative epistemology. Interdisciplinarity is observed when two or more disciplines establish relational bonds to achieve diversified and broader knowledge about a given phenomenon [36].

The qualitative approach of the study allows an approximation of the research objects [37], enabling the understanding and explanation of phenomena to subsequently be used in the statistical research [38]. The search for relationships between different disciplines can be performed by the existing Systematic Literature Review (SLR) [39]. SLR can be performed in two formats: (i) the first format focuses on an already established and mature literature in which the body of knowledge needs analysis and synthesis; and (ii) a second format involves an emerging theme whose exposure would benefit potential theoretical foundations. This study focuses on the second format of SLR by discussing the relationship between themes (e-Government and Green IT) that are emerging in the scientific literature.

For an SLR to be considered reliable, Webster and Watson [39] suggest that the main contributions are probably in the main scientific journals. Thus, we conducted a systematic search in the main scientific databases that index the main journals, according to Table 5.1. In addition to the application of search strings, the abstracts of the studies were read to verify if there was any kind of relationship between the themes e-Government and Green IT.

Table 5.1, only five studies were considered after the inclusion criteria application. To expand the search for studies that had the themes e-Government and Green IT together, the search strings were changed to consider one of the two terms in the abstract and the other in the body of the article. This strategy was adopted alternately in the same databases, according to Table 5.2

Through the search strings of Tables 5.1 and 5.2, a total of 62 articles were considered as a result of SLR. After the identification of studies in the main academic databases, we sought the identification of seminal and derivative studies with close alignment to the researched phenomenon. To achieve this goal, we used the study by Sedera et al. [9] on the ConnectedPapers platform. The ConnectedPapers

Table 5.1 Review of academic databases with e-Government and Green IT themes together

Database	Search string	Results	Excluded	Included
SCOPUS	(TITLE-ABS-KEY (e-Government) AND TITLE-ABS-KEY ("Green IT"))	15	12	3
Web of science	e-Government (Topic) AND "Green IT" (Topic)	1	0	1
EBSCOhost	TX "Green IT" AND TX e-Government	11	11	0
Springerlink	With exact phrase: Green IT AND where the titles contains: e-government	3	2	1
	With exact phrase: e-government AND where the titles contains: Green IT	3	3	0
Springer Nature	Title: e-government	5	5	0
	Title: "Green IT"	0	0	0
Grand total		38	33	5

Table 5.2 Review in academic databases with e-Government and *Green IT* themes in the body of studies

Database	Search string	Results	Excluded	Included
SCOPUS	(TITLE-ABS-KEY (e-government) AND ALL ("Green IT")))	47	11	36
SCOPUS	(TITLE-ABS-KEY ("Green IT") AND ALL (e-government))	32	1st 1	21
Wos	e-government (Topic) AND "Green IT" (All fields)	1	1	0
Wos	"Green IT" (Topic) AND e-government (All fields)	1	1	0
Grand total		81	24	57

(https://www.connectedpapers.com/) platform is a visual tool that helps scientists and researchers find and explore relevant work in their fields of research [40]. Through the ConnectedPapers platform, the association graph was generated as shown in Fig. 5.1. Sedera et al. [9] were chosen because it is a work that observes the theme "Green IT" from a multidisciplinary perspective and presents ways for future studies on the subject, including e-Government.

Sedera's et al. [9] study with a larger circle in the center of the Fig. 5.1. Although there is no direct association of Sedera's et al. [9] study with the other studies presented in the graph, the graph suggests the existence of co-citations and bibliographic coupling [40] with Sedera's et al. [9] work, suggesting that they are studies with similar approaches. Following, also through the ConnectedPapers platform, the studies suggested as seminal are listed according to the work of Sedera [9]—Table 5.3.

Table 5.3 those studies with a high number of citations and which, because of their relevance in the literature, tend to be seminal studies within the central theme of the base document—the scientific article published by Sedera et al. [9]. Next, Table 5.4 presents the ten studies considered by the ConnectedPapers platform as derivatives, that is, studies that have a close relationship to the work of Sedera et al. [9].

Considering that the previous analyses are based only on the theme "Green IT", we performed the same procedure for a study that has as its central theme the "e-Government". The study chosen is titled "e-government service research development: A literature review" by Hasan [15]. The criterion of the choice of the study was given by its type—Systematic Literature Review. SLRs tend to present the main works of the researched literature and may have a better adaptation to tools such as ConnectedPapers. Following, the ConnectedPapers platform graph for Hasan's study [15] is presented.

Hasan's study [15] is presented in Fig. 5.2, in the center of the figure. Hasan's work [15] has connections with several other works in the literature that address the theme "e-Government". To highlight these studies, we present below the studies that are considered seminal on the theme—Table 5.5.

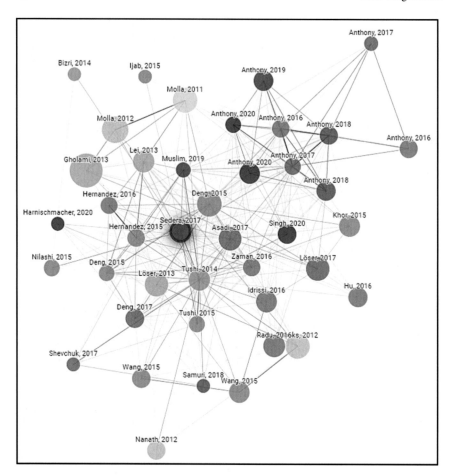

Fig. 5.1 ConnectedPapers (Green IT) platform association graph. (Developed by ConnectedPapers platform based on Sedera et al. [9])

Table 5.3 Prior works according to Sedera et al. [9]

Paper	Author	Year	Number of citations	Number of references on graph
[41]	Lindsay McShane	2011	394	23
[7]	T. Butler	2011	230	21
[6]	S. Murugesan	2008	783	25
[42]	Nigel P. Melville	2010	1032	32
[30]	Adela J. Chen	2010	998	33

Adapted from ConnectedPapers platform based on Sedera [9]. The complete table can be accessed at: https://www.connectedpapers.com/main/f7eb78909526a574c609893d4e3958b99226b243/Multi%20disciplinary-Green-IT-Archival-Analysis%3A-A-Pathway-for-Future-Studies/prior

Table 5.4 Derivative works according to Sedera et al. [9]

Paper	Author	Year	Number of citations	Number of references on graph
[43]	Helge Schmermbeck	2019	6	10
[11]	Muhammad Ashraf Fauzi	2020	23	7
[10]	Bokolo Anthony Jnr	2020	31	7
[44]	Romli Awanis	2017	7	7

Adapted from ConnectedPapers platform based on Sedera et al. [9]. The complete table can be accessed at: https://www.connectedpapers.com/main/f7eb78909526a574c609893d4e3958b99226 b243/Multi%20disciplinary-Green-IT-Archival-Analysis%3A-A-Pathway-for-Future-Studies/der ivative

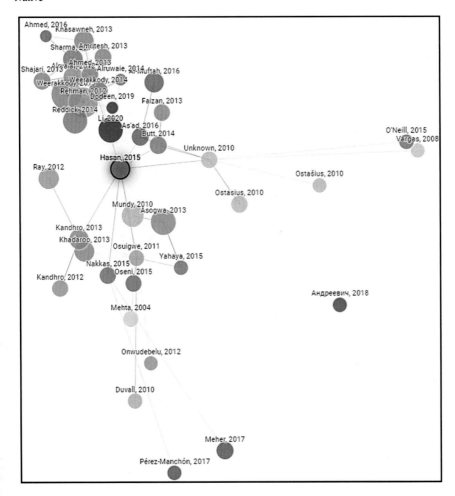

Fig. 5.2 ConnectedPapers e-Government platform association graph. (Developed by Connected-Papers platform based on Hasan [15])

Table 5.5 Prior works according to Hasan [15]

Paper	Author	Year	Number of citations	Number of references on graph
[21]	Yogesh Kumar Dwivedi	2011	470	12
[20]	Jungwoo Lee	2001	2442	10
[22]	F. Bélanger	2005	1875	10
[45]	Fred D. Davis	1989	41,348	9
[46]	Fred D. Davis	2003	25,234	8

Adapted from ConnectedPapers platform based on Hasan [15]. The complete table can be accessed at: https://www.connectedpapers.com/main/5f8ba4b020e7eb2aad58d97aac717b0eaf93c00e/E%20Government-Service-Research-Development%3A-A-Literature-Review/prior

Table 5.6 Derivative works according to Hasan [15]

Paper	Author	Year	Number of citations	Number of references on graph
[48]	P. Hart	2015	10	7
[49]	P. Hart	2015	2	6
[50]	Aynur Gurbanli	2018	3	5
[16]	M. Reuver	2018	110	3
[51]	H. Ranaweera	2016	34	3

Adapted from ConnectedPapers platform based on Hasan [15]. The complete table can be accessed at: https://www.connectedpapers.com/main/5f8ba4b020e7eb2aad58d97aac717b0eaf93c00e/E%20Government-Service-Research-Development%3A-A-Literature-Review/derivative

The seminal papers presented in Table 5.5 use of theoretical bases in the literature in the field of Information Systems (e.g. Davis et al. [45]; DeLone and McLean [47]; Venkatesh et al. [46]) to discuss the topic of e-Government. Most of this literature analyzes the acceptance of technologies and information systems by users. Next, in Table 5.6, we present the derivative works according to Hasan [15].

The studies presented in the central theme the "e-Services" together with "e-Government", may suggest an alignment between these two concepts in recent studies.

After identifying the main studies related to the studied phenomenon and its theoretical foundations (seminal studies), we used the software for managing bibliographic references Mendeley (https://www.mendeley.com/) to index the selected studies. In the sequence, we used the Rayyan platform (https://rayyan.ai/) to identify the bibliometric data of the 102 publications (62 of the SLR + 40 of the Connected-Papers). Ryyan is a free online application that assists researchers with systematic literature reviews [52].

Iramuteq version 0.7 alpha 2 software was used for the analysis of the publications. The Iramuteq software presents different lexical analysis techniques that can be applied to the textual corpus [53]. Iramuteq allows the realization of the Descending Hierarchical Classification (DHC) method to classify text segments about their respective vocabularies. Vocabulary is separated based on the frequency of reduced shapes (words already stemmed). The objective of this analysis is to identify classes of Elementary Context Units (UCE) or Text Segments (ST). The UCE presents similar vocabularies of a class and distinct vocabularies between the classes [54]. Iramuteq also allows the performance of Factorial Correspondence Analysis (FCA) from CHD (post-factor analysis). Through a Cartesian plane, the different words and variables associated with each of the DHC classes are presented. In this sense, the analysis allows obtaining the text segments associated with each class, enabling a more qualitative analysis of the data [54].

The Iramuteq software also allows the analysis of the same and the cloud of words to be performed. The analysis of similitude is based on graph theory and presents the connection between words. The word cloud is one of the simplest but most interesting analyses because it allows a graphic representation of the words more frequently in the analyzed corpus [53, 54]. The textual corpus used in the Iramuteq software was extracted from the abstracts of the 102 studies selected in the SLR.

For the execution of the analyses via Iramuteq software the texts extracted from the abstracts went through a process of standardization due to the diversity of forms in which the same term can be written. The terms "eService", "Eservice" or "eservice" have been replaced by the word "e-service". The terms "e-Government", "e-government" and "E-government" were replaced by "and government". The terms "green-IT", "Green IT" and "Green IT" have been replaced by "green_it".

5.5 Analysis and Discussions

The first analysis performed refers to Zipf's Bibliometric Law. Zipf's law suggests that the most commonly used words indicate the subject of a document [55]. To observe the data according to Zipf's law, the word cloud was created for the abstracts of the 102 studies identified in the SLR (see Fig. 5.3).

The word cloud highlights the word "egovernment" with a frequency equal to 193. Following are the words "service" (136), "research" (134), "study" (132), "model" (120), "information" (117), "technology" (101), "government" (97), "envinronmental" (93) and "sustanability" (93). The other words had a frequency of less than 90. The term "green_IT" presented a frequency equal to 68, suggesting that even if the search for studies in which the term is central, this term ends up being applied in a smaller proportion, predominantly terms such as "envinronmental" (93) and "sustanability" (93).

Next, in Table 5.7, the Bradford Bibliometric Law was applied, which estimates the degree of relevance of journals in a given knowledge field [55]. Among the 102 studies identified, 62 are articles published in scientific journals, 34 are articles published in scientific events, 4 are books, 2 are chapters, and one doctoral thesis.

The scientific journal with the largest number of publications on the theme "e-Government" is "Government Information Quarterly—ISSN: 0740-624X" with six publications. Following is observed the scientific journal MIS Quarterly—ISSN:

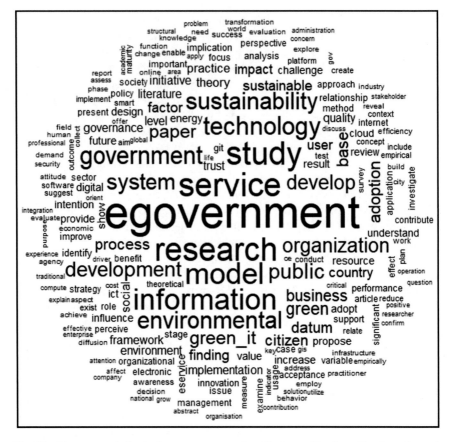

Fig. 5.3 Abstract word cloud—frequency greater than 10. *Source* Developed by the authors via Iramuteq software

Table 5.7 List of journals or scientific events with the largest number of publications (value above 1)

Title of the journal or scientific event	Number of publications
Government information quarterly	9
MIS quarterly	4
Journal of strategic information systems	4
15th Pacific Asia conference on information systems: quality research in pacific, PACIS 2011	2
Journal of theoretical and applied information technology	2
Commun. Assoc. Inf. Syst	2
32nd international conference on information system 2011, ICIS 2011	2

Table 5.8 List of more prolific authors on Green IT and e-Government

Author's name	Number of publications
Deng, H	3
Pan, S L	3
Zuo, M	3
Karunasena, K	3

0276–7783 and EISSN: 2162–9730 with four publications. The journal MIS Quarterly, unlike the journal Government Information Quarterly, presents publications more directed to topics associated with Technology Acceptance [56] and, Information Systems and Environmental Sustainability [30, 31, 42]. The third journal with the most publications presented at SLR was the Journal of Strategic Information Systems—ISSN: 9,638,687—with four publications. This journal presents studies with themes similar to MIS Quarterly journal, such as the proposition of a framework to integrate Information Technology with sustainable practices [8, 32], Information Systems, and Ecological Sustainability [12] or a theory for "Green IS" [7].

Next, Table 5.8 List of more prolific authors on Green IT and e-Government—presents the list of authors with the largest number of publications according to the SLR is presented.

The authors Deng, H (College of Business and Law—RMIT University, Australia), Gurbanli, Aynur (Institute of Information Technology, Azerbaijan), Pan S L (University of New South Wales, Australia), Zuo, M (Renmin University of China, China) and Karunasena, K (School of Business Information Technology and Logistics, RMIT University) presented the highest number of publications (3). Other authors presented two publications as Hart, P (School of Computing, Faculty of Technology, University of Portsmouth, United Kingdom), Chen, Adela J (University of Georgia, Department of Management Information Systems), Boudreau, Marie-Claude (Department of Management Information Systems, Terry College Business—University of Georgia), Helge Schmermbeck (University of Duisburg-Essen), among other researchers, had the highest number of publications.

It is observed by the affiliation of the most prolific authors a predominance of researchers originating from the Information Systems field. The analyses suggest that both discussions involving e-government and Green IT are focused on research groups in the field of Information Technology/Information Systems.

Next, in Fig. 5.4, we observed how the publications of the two themes occurred over the years according to our SLR.

The largest number of publications (18) occurred in 2011, followed by 2019 (10). Figure 5.4 shows that studies involving e-Government and Green IT have gained more attention from researchers since 2008. Since then, the number of publications has maintained an average of 7 publications per year, suggesting that the themes, together, are still little researched.

To complement the analyses and understand how the e-Government and Green IT themes are related in the studies selected by the SLR, we present the Similitude Analysis according to the abstract texts (see Fig. 5.5).

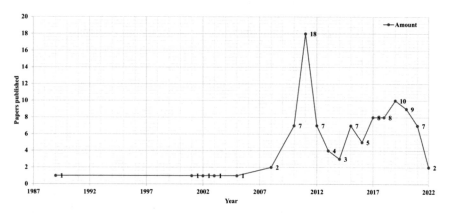

Fig. 5.4 Publications over the years

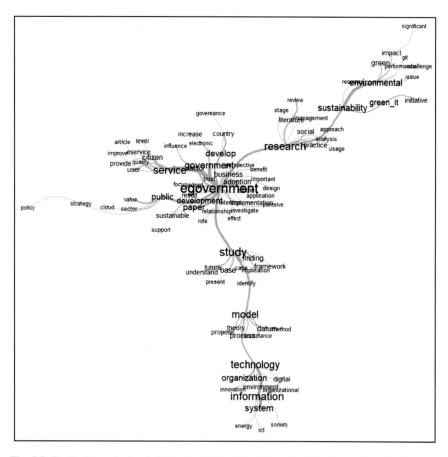

Fig. 5.5 Similarity analysis of abstracts—102 articles. (Developed by the authors via Iramuteq software)

Similarity Analysis allows us to observe that the theme "e-Government" is central in the studies and that the discussions associated with sustainability, environment, and Green IT are secondary themes that branch out from discussions on e-Government. In another branch of Fig. 5.5, it is possible to observe that the themes of Information Technology and Information Systems are also part of secondary discussions associated with e-Government.

Although the analysis of similarities allows understanding how words most frequently in the text are related, the analysis does not allow observing these words through similar lexical classes or groups. Thus, we performed the analysis by Reinert's Method—Descendent Hierarchical Classification (DHC)—see Fig. 5.6.

Analysis using the Reinert Method presented four classes, and classes 2 (28.5%) and 3 (27.7%) presented the highest percentages in the volume of texts in the class. Next, in Fig. 5.7, we can see the classes with their main words.

According to Fig. 5.7 is possible to observe that class 1 (red) has words associated with the e-Government theme such as "government", "public", "service" and "eservice". The following excerpts are from class 1 (red):

> e-Government services adoption rate is rapidly increasing in the developing and lower-middle-income countries for promoting good governance capability and accountability of public organisations and government services are helping to boost government revenue very fast and secured transactions.

> The rapid diffusion of internet couple with the digitalization of most economies in the world today which have given rise to egovernment services adoption to promote good governance capability and accountability of public organisations.

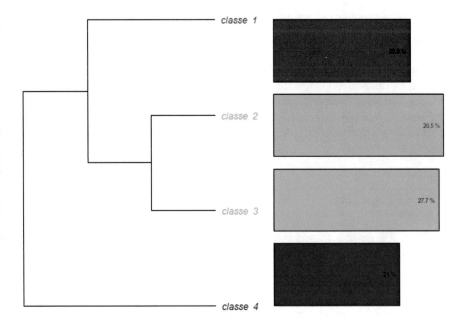

Fig. 5.6 Descending hierarchical classification. (Developed by the authors via Iramuteq software)

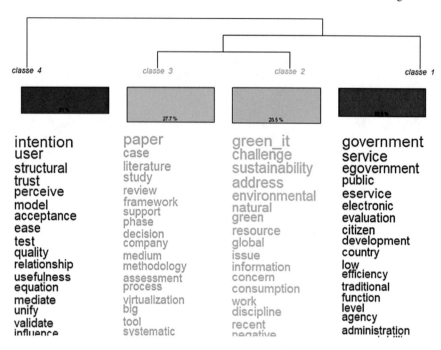

Fig. 5.7 DHC with the main words (Developed by the authors via Iramuteq software)

There are still concerns about eservice security in developing countries as consumers and online users have become the victims of coordinated cybercrimes despite the fact that egovernment services are helping to boost government revenue through very fast transactions reduce corruption through the use of modern technology and transparent operations.

Sections of Class 1 suggest that e-Government can promote better governance of governments and help increase revenue because it enables faster transactions and reduces corruption through the use of technology that enables transparent operations.

Class 2 (green) features words associated with the Green IT theme such as "challenge", "sustainability", "environmental", "natural" and "green"—see Class 2 excerpts:

...green-friendly policies and practices sustainability is a significant field in both research and practice as such collaborative enterprise are adopting sustainable initiatives to achieve efficient resource usage, CO_2 emissions reduction, moderate cost incurred, ethical waste management, and natural resources conservation.

...green it represents a dramatic change in priority in the IT industry so far the industry has been focusing on IT equipment processing power and associated equipment spending it's not been concerned with other requirements such as power.

...practical implication for this study provides empirical evidence from IT executives emphasizing that green IS is capable of decreasing the environmental effects of traditional IT infrastructure deployment in organizations, there is a considerable amount of awareness of environmental issues and corporate responsibility for sustainability.

Class 2 excerpts suggest that studies involving Green IT are more associated with research involving industry/organizations and their IT infrastructure. No studies discuss the IT infrastructure of governments and its relationship with the concept of Green IT.

Class 3 (lilac) presents words associated with generic terms of scientific articles such as "paper", "case", literature", "review", and "methodology" among others. Finally, class 4 (purple) presents words associated with the Technology Acceptance literature [45] such as "intention", "use", "trust", "perceive", "usefulness" and "acceptance". Moreover, it is observed in class 4 words that suggest the type of analysis predominant in studies such as "structural" and "equation"—see Class 4 excerpts:

Through the review of prior green IS and Green IT studies, this article proposes a Green IT structural model based on the unified theory of acceptance and use of technology after which survey data were collected from IT executives in various organizations in Malaysia.

The models include the Delone and Mclean model and Trust theory which involves eight variables: system information quality, information quality, service quality, trust in government organizations, trust in technology, usage, user satisfaction, net benefits, and sustainable information society.

Class 4 excerpts highlight the use of models and theories in the Information Systems field as the theoretical basis of studies involving both Green IT and e-Government.

Following, Fig. 5.8, the Factor Analysis of Correspondence (FAC) is presented. FAC represents, in a Cartesian plane, the different words and variables associated with each of the DHC classes.

FAC allows us to observe in a Cartesian plan the four classes identified by DHC and their lexical proximities [53]. Through Fig. 5.8 it is possible to observe that the textual corpus referring to e-government (Class 1—red) and Green IT (Class 2—green) present lexical distances, including in the quadrants in which they are allocated in the Cartesian plane. These analyses allow us to infer that the themes do not present lexical association in the studied literature (RSL result). Thus, we conclude that the literature on e-Government, despite being theoretically discussed with foundations from the Information Systems literature, has not been analyzed in conjunction with the literature on Green IT.

5.6 Discussions

It is important at this point to revisit the objective of the study: to identify the intersection point between the scientific literature on e-Government and Green IT to enable interdisciplinary research lines involving these subjects. Our analyses suggest that e-Government and Green IT subjects have no associations in the literature studied. Each of the themes is discussed and analyzed differently, but they share similar theoretical underpinnings in the Information Systems field.

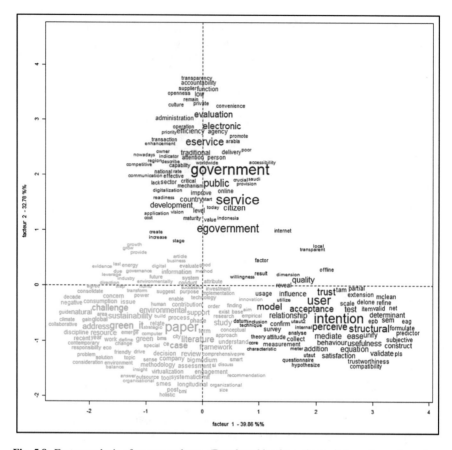

Fig. 5.8 Factor analysis of correspondence. (Developed by the authors via Iramuteq software)

However, although we were not able to identify a point of intersection between e-Government and Green IT subjects, our study presents interesting results that may guide future studies. We have identified the main scientific journals about e-Government (Government Information Quarterly), and Green IT (MIS Quarterly and Journal of Strategic Information Systems).

Furthermore, we observed that studies involving the theme of e-Government have been theoretically based on the classical Technology Acceptance Models (e.g., TAM, UTAUT, and UTAUT2) and the application of statistical methods by Structural Equation Modeling. Otherwise, the literature on Green IT still lacks a more robust theoretical foundation, especially if observed in the context of governments. We observe that there is a predominance of studies on Green IT, but with a focus on organizations/industries rather than governments.

5.7 Conclusions

We can conclude that e-Gov theory grew fast since its first propositions in the middle 90 s, with a surge of works in the following decade, but focused on the technical aspect, without considering social and political nuances that permeate public administration and management. Although widely adopted and fast-growing, research on the subject lack a more solid theoretical ground, especially due to the complexity of the e-Gov construct. However, this scenario is getting better as more researchers explore other aspects beyond the ICT approach. The field continues to evolve following ICT technological advances and many dedicated conferences and journals provide ample access for researchers to produce new work.

Regarding Green IT literature, despite much focus and research efforts culminating in the emergence of many theoretical frameworks, there is still much space for realizing those theories into actual practices. A considerable body of research helped to consolidate Green IT and Green IS as valuable environmental sustainability driving forces. Nevertheless, there is a lack of convergence with e-Gov initiatives, though the recent migration of those services to cloud-based solutions may have contributed to improving the scenario of green practices in e-Gov. Prospects to further integrate Green IT and Green IS practices into e-Gov adoption models and implementation frameworks present a considerable source for future research.

Finally, we consider that the duality of e-Government and Green IT is especially suitable for governments who intend to develop the e-Government concept while following sustainable environmental politics. Thus, we suggest that new studies try to answer questions such as: How to minimize the environmental impact of ICTs at the same proportion that governments become more electronic (e-Government)? Does the fact that the government develops Green IT practices influence the application of these same practices in the country's industry and/or organizations? Should Green IT practices be initially developed in industries and organizations, or in the government itself? Can governments be a model for the application of Green IT principles to their industries and organizations? How aware are governments, industries, and organizations of the Green IT concept?

Even when methodological practices inherent to a scientific study are applied, limitations must be considered. The inclusion criteria for the Systematic Literature Review were based on abstract reading and researchers' analysis of the adequacy or not of the paper for the research objective. Only abstract reading does not represent the study's overall context. Furthermore, qualitative analysis of the appropriateness or not of a study for the SLR may be influenced by the researchers' prior knowledge of the subjects under observation.

References

1. Liang, Y., Qi, G., Zhang, X., & Li, G. (2019). The effects of e-Government cloud assimilation on public value creation: An empirical study of China. *Government Information Quarterly, 36*. https://doi.org/10.1016/j.giq.2019.101397

2. Heeks, R., & Bailur, S. (2007). Analyzing e-government research: Perspectives, philosophies, theories, methods, and practice. *Government Information Quarterly, 24*, 243–265. https://doi.org/10.1016/j.giq.2006.06.005
3. Deng, H. (2008). Towards objective benchmarking of electronic government: An inter-country analysis. *Transforming Government People, Process Policy, 2*, 162–176. https://doi.org/10.1108/17506160810902176
4. Antoni, D., Apriliani, Herdiansyah, M.I., & Akbar, M. (2018). Critical factors of transparency and trust for evaluating e-government services for the poor. In *2nd International Conference on Informatics and Computing, ICIC 2017*, pp. 1–6. Institute of Electrical and Electronics Engineers Inc., Fakultas Ilmu Komputer, Universitas Bina Darma.
5. Krishnan, S., & Teo, T. S. H. (2011). e-Government, e-business, and national environmental performance. In *PACIS 2011—15th Pacific Asia Conference on Information Systems: Quality Research in Pacific*. Department of Information Systems, School of Computing, National University of Singapore.
6. Murugesan, S., & Gangadharan, G. R. (2012). *Harnessing Green IT: Principles and Practices*. Wiley-IEEE Computer Society PR
7. Butler, T. (2011). Compliance with institutional imperatives on environmental sustainability: Building theory on the role of Green IS. *The Journal of Strategic Information Systems, 20*, 6–26. https://doi.org/10.1016/j.jsis.2010.09.006
8. Dao, V. T., Langella, I. M., & Carbo, J. (2011). From green to sustainability: Information technology and an integrated sustainability framework. *The Journal of Strategic Information Systems, 20*, 63–79. https://doi.org/10.1016/j.jsis.2011.01.002
9. Sedera, D., Lokuge, S., Tushi, B., & Tan, F. (2017). Multi-disciplinary green IT archival analysis: A pathway for future studies. In: *Communications of the Association for Information Systems*, p 28. Association for Information Systems.
10. Jnr, B. A. (2020). Examining the role of green IT/IS innovation in collaborative enterprise-implications in an emerging economy. *Technology in Society, 62*, 101301. https://doi.org/10.1016/j.techsoc.2020.101301
11. Ojo, A. O., & Fauzi, M. A. (2020). Environmental awareness and leadership commitment as determinants of IT professionals engagement in Green IT practices for environmental performance. *Sustainable Production and Consumption*. https://doi.org/10.1016/j.spc.2020.07.017
12. Chen, A. J., Boudreau, M.-C., & Watson, R. (2008). Information systems and ecological sustainability. *The Journal of Strategic Information Systems, 10*, 186–201. https://doi.org/10.1108/13287260810916907
13. Houghton, J. (2007). ICT and the environment: A framework for analysis. http://www.oecd.org/sti/ieconomy/40833025.pdf
14. Kumar, R., & Mieritz, L. (2007). *Conceptualizing "Green" IT and Data Center Power and Cooling Issues*
15. Lange, S., Pohl, J., & Santarius, T. (2020). Digitalization and energy consumption. Does ICT reduce energy demand? *Ecological Economics, 176*, 106760. https://doi.org/10.1016/j.ecolecon.2020.106760
16. Bouwman, H., Nikou, S., Molina-Castillo, F.J., & Reuver, M. (2018). The impact of digitalization on business models. *Digital Policy, Regulation and Governance, 20*, 105–124. https://doi.org/10.1108/DPRG-07-2017-0039
17. Mahmudul Hasan, M. (2015). e-Government service research development: A literature review. *Public Affairs and Administration Concepts, Methodologies, Tools and Application, 1*, 538–568. https://doi.org/10.4018/978-1-4666-8358-7.CH025
18. Lee, J., Kim, H. J., & Ahn, M. (2011). The willingness of e-Government service adoption by business users: The role of offline service quality and trust in technology. *Government Information Quarterly, 28*, 222–230. https://doi.org/10.1016/j.giq.2010.07.007
19. Lips, A. M. B., & Schuppan, T. (2009). Transforming e-government knowledge through public management research. *Public Management Review, 11*, 739–749. https://doi.org/10.1080/14719030903318921

20. Layne, K., & Lee, J. (2001). Developing fully functional e-government: A four stage model. *Government Information Quarterly, 18*, 122–136. https://doi.org/10.1016/S0740-624X(01)000 66-1
21. Shareef, M., Kumar, V., Kumar, U., & Dwivedi, Y. K. (2011). e-Government adoption model (GAM): Differing service maturity levels. *Government Information Quarterly, 28*, 17–35. https://doi.org/10.1016/j.giq.2010.05.006
22. Carter, L. D., & Bélanger, F. (2005). The utilization of e-government services: citizen trust, innovation and acceptance factors. *Information System Journal, 15*. https://doi.org/10.1111/j. 1365-2575.2005.00183.x
23. Welch, E. W., Hinnant, C. C., & Moon, M. J. (2005). Linking citizen satisfaction with e-government and trust in government. *Journal of Public Administration Research and Theory, 15*, 371–391. https://doi.org/10.1093/jopart/mui021
24. Papadomichelaki, X., & Mentzas, G. (2012). e-GovQual: A multiple-item scale for assessing e-government service quality. *Government Information Quarterly, 29*, 98–109. https://doi.org/ 10.1016/j.giq.2011.08.011
25. Wyld, D. C. (2010). The cloudy future of government it: cloud computing and the public sector around the world. *International Journal of Web and Semantic Technology, 1*, 1–21.
26. Ingrams, A., Manoharan, A., Schmidthuber, L., & Holzer, M. (2020). Stages and determinants of e-government development: A twelve-year longitudinal study of global cities. *International Public Management Journal, 23*, 731–769. https://doi.org/10.1080/10967494.2018.1467987
27. Malodia, S., Dhir, A., Mishra, M., & Bhatti, Z. A. (2021). Future of e-Government: An integrated conceptual framework. *Technological Forecasting and Social Change, 173*, 121102. https://doi.org/10.1016/j.techfore.2021.121102
28. Saylam, A., & Yıldız, M. (2022). Conceptualizing citizen-to-citizen (C2C) interactions within the E-government domain. *Government Information Quarterly, 39*. https://doi.org/10.1016/j. giq.2021.101655
29. Umbach, G., & Tkalec, I. (2022). Evaluating e-governance through e-government: Practices and challenges of assessing the digitalisation of public governmental services. *Evaluation and Program Planning, 93*, 102118. https://doi.org/10.1016/j.evalprogplan.2022.102118
30. Watson, R., Boudreau, M.-C., & Chen, A. J. (2010). Information systems and environmentally sustainable development: Energy informatics and new directions for the IS community. *MIS Quarterly, 34*, 23–38. https://doi.org/10.2307/20721413
31. Elliot, S. (2011). transdisciplinary perspectives on environmental sustainability: A resource base and framework for IT-enabled business transformation. *MIS Quarterly, 35*, 197–236. https://doi.org/10.2307/23043495
32. Bose, R., & Luo, X. (2011). Integrative framework for assessing firms' potential to undertake Green IT initiatives via virtualization—A theoretical perspective. *The Journal of Strategic Information Systems, 20*, 38–54. https://doi.org/10.1016/j.jsis.2011.01.003
33. Landum, M., Moura, M. M. M. E, & Reis, L. (2021). A framework for the Alignment of ICT with Green IT. *Advance in Science, Technology and Engineering System Journal, 6*, 593–601. https://doi.org/10.25046/aj060268
34. Verdecchia, R., Lago, P., Ebert, C., & De Vries, C. (2021). Green IT and green software. *IEEE Software, 38*, 7–15. https://doi.org/10.1109/MS.2021.3102254
35. Anthony, B., Majid, M. A., & Romli, A. (2020). A generic study on Green IT/IS practice development in collaborative enterprise: Insights from a developing country. *Journal of Engineering and Technological Management—JET-M, 55*, 101555. https://doi.org/10.1016/j.jengte cman.2020.101555
36. Bernstein, J. H. (2014). Disciplinarity and trandisciplinarity in the study of knowledge. *Informing Science, 17*, 241–273.
37. Minayo, M.C. de S., & Sanches, O. (1993). Quantitativo-qualitativo: oposição ou complementaridade? *Cadernos de Saude Publica, 9*, 237–248. https://doi.org/10.1590/S0102-311X19930 00300002
38. Minayo, M.C. de S. (2012). Análise qualitativa: teoria, passos e fidedignidade. *Ciência and Saúde Coletiva, 17*, 10. https://doi.org/10.1590/S1413-81232012000300007

39. Webster, J., & Watson, R. T. (2002). Analyzing the past to prepare for the future: Writing a Review. *MIS Quarterly, 26*, xiii–xxiii. https://doi.org/10.2307/4132319
40. Eitan, A. T., Smolyansky, E., & Harpaz, I. K., Connected Papers. https://www.connectedpapers.com/about
41. Jenkin, T. A., Webster, J., & McShane, L. (2011). An agenda for "Green" information technology and systems research. *Information and Organization, 21*, 17–40. https://doi.org/10.1016/j.infoandorg.2010.09.003
42. Melville, N. P. (2010). Information systems innovation for environmental sustainability. *MIS Quarterly, 34*, 1–21. https://doi.org/10.2307/20721412
43. Schmermbeck, H. (2019). On making a difference: Towards an integrative framework for green IT and green IS adoption. In *Hawaii International Conference on System Sciences*. Hawai, EUA.
44. Jnr, B. A., Mazlina, A. M., & Awanis, R. (2017). Sustainable adoption and implementation in collaborative enterprise: A systematic literature review. *Journal of Theoretical and Applied Information Technology, 95*, 1875–1915.
45. Davis, F. D. (1989). Perceived usefulness, perceived ease of use, and user acceptance of information technology. *MIS Quarterly, 13*, 319–340. https://doi.org/10.2307/249008
46. Venkatesh, V., Morris, M. G., Davis, G. B., & Davis, F. D. (2003). User acceptance of information technology: Toward a unified view. *MIS Quarterly, 27*, 425–478. https://doi.org/10.2307/30036540
47. DeLone, W. H., & McLean, E. R. (2003). The DeLone and McLean model of information systems success: A ten-year update. *Journal of Management Information Systems, 19*, 9–30. https://doi.org/10.1080/07421222.2003.11045748
48. Oseni, K. O., Dingley, K., & Hart, P. (2015). Barriers facing e-Service technology in developing countries: A structured literature review with Nigeria as a case study. In *2015 International Conference on Information Society (i-Society)*, pp. 97–103.
49. Oseni, K. O., Dingley, K., & Hart, P.(2015). e-Service security: taking proactive measures to guide against theft, case study of developing countries. *International Journal of e-Learning, 5*, 454–461. https://doi.org/10.20533/IJELS.2046.4568.2015.0058
50. Yusifov, F., & Gurbanli, A. (2018). e-Services evaluation criteria: the case of Azerbaijan. Information and Moksl. https://doi.org/10.15388/IM.2018.0.11938
51. Ranaweera, H. (2016). Perspective of trust towards e-government initiatives in Sri Lanka. *Springerplus, 5*. https://doi.org/10.1186/s40064-015-1650-y
52. Johnson, N., & Phillips, M. (2018). Rayyan for systematic reviews. *Journal of Electronic Resources Librarianship, 30*, 46–48. https://doi.org/10.1080/1941126X.2018.1444339
53. Marchand, P., & Ratinaud, P. (2012). L'analyse de similitude appliquée aux corpus textuels : les primaires socialistes pour l'élection présidentielle française. In *Actes des 11èmes Journées Internationales d'Analyse des Données Textuelles (JADT)*, pp. 687–699.
54. Camargo, B. V., & Justo, A. M. (2013). IRAMUTEQ: Um software gratuito para análise de dados textuais. *Temas em Psicologia, 21*, 513–518. https://doi.org/10.9788/TP2013.2-16
55. Frogeri, R. F., Ziviani, F., Martins, A. de P., Maria, T. C., & Zocal, R. M. F. (2022). O Grupo de trabalho 4 do enancib: uma análise bibliométrica. *Perspect. em Gestão Conhecimento, 12*, 235–252. https://doi.org/10.22478/ufpb.2236-417X.2022v12n1.62824
56. Davis, F. D., Bagozzi, R. P., & Warshaw, P. R. (1989). Perceived usefulness, perceived ease of use, and user acceptance of information technology. *MIS Quarterly, 27*, 319–340. https://doi.org/10.1016/S0305-0483(98)00028-0

Chapter 6
Machine Learning Technique for Predicting the Rural Citizens' Trust on Using e-Governance Health Care Applications During COVID-19

M. Bhuvana, A. Ramkumar, and B. Neeraja

Abstract Modernization in ICT (Information and Communication Technology) provides an outstanding contribution in offering health care services across the globe. In today's world the communication between the citizens and the government takes place through various innovative technologies and these advances in ICT has restructured the working culture of the citizens residing in our country. The phrase "e-governance" has become a pivotal tool in the twenty-first century to develop applications for facilitating services to the citizens (United Nations [UN] E-Government Survey 2018). An epidemic deadly virus called "COVID-19" (Coronavirus 2019) was outlined by WHO (World Health Organization) in December 2019 at Wuhan city, China. This deadly virus has been infected by the entire world and various initiatives have been taken by the government to safeguard the peoples' life. In developing countries like India, out of total population, 70% of the people residing in remote villages. Multiple challenges have been faced by the government officials in reaching and delivering their services to the rural people. The Present research study has focused on predicting the actual dimension of trust among rural citizens towards accessing applications for e-governance health care services. In this present study, the author has used "DeLone and McLean 2003 Information System Model" for constructing the theoretical framework. The researchers have predicted that "Access to information is the exact dimension that is strongly associated with the outcome variable "Rural Citizens' Trust" on accessing e-governance health care applications through Neural Network analysis in R studio.

Keywords COVID-19 · e-Governance health care applications · Rural Citizens' Trust

M. Bhuvana (✉) · A. Ramkumar
Department of Business Administration, Vels Institute of Science, Technology and Advanced Studies (VISTAS), Chennai, Tamilnadu, India
e-mail: bhuvana.sms@velsuniv.ac.in

A. Ramkumar
e-mail: ram.sms@velsuniv.ac.in

B. Neeraja
Expert Solutions Technologies (EST) Canada, Inc, Chennai, Tamilnadu, India

C. Gaie and M. Mehta (eds.), *Recent Advances in Data and Algorithms for e-Government*,
Artificial Intelligence-Enhanced Software and Systems Engineering 5,
https://doi.org/10.1007/978-3-031-22408-9_6

6.1 Introduction

The term "E-governance" is defined as the process of providing access to Information and Communication Technology (ICT) based governmental services for the citizens in the country. It is determined as the most convenient mode for the citizen and government to communicate and sharing of information [46]. It has been identified as the essential force for the continuous improvement in the efficiency, effectiveness and quality of the governance [4]. Awareness, Literacy Level, Users' acceptability, Legal and Digitalization are the major sub areas of e-governance that has to be focused on transformational improvement in the quality of the government [42].

E-governance is described as the term "SMART" that represents a "Simple", "Moral", "Accountable", "Responsible", and "Transparent" method of interacting with the people residing in the country. The primary objective of e-governance services is to deliver its services to the people who are residing in remote villages in the society [2]. According to the report of World Bank 2018, in India, about 70% of the entire population engaged with the rural population. Hence the government services should reach the rural people at their doorstep to a greater extent [24]. To satisfy the immediate and fundamental requirements of the people, the Government of India (GOI) has taken several initiatives in different sectors like Agriculture, Education, Health Care, Insurance, Banking services, and other benefits such as renewal of identification proofs, payment of bills, and taxes, etc. For satisfying these objectives, the Government of India (GOI) has structured common service centers (CSCs) at peoples' localities across the states [5–7].

A new pneumonia virus "Novel CORONA Virus 2019" also called "COVID-19" has been recognized in December 2019 by the WHO (World Health Organization) in China. This virus directly attacks the respiratory system of human beings [52]. There are millions of people have been infected by this deadly virus even in the most developed countries like Brazil, Italy, the United States, Italy, Germany, and United Kingdom (UK) (Civil Service India, 2020). For more than 10 months the entire world has been in complete lockdown. The World Health Organization (WHO) has directed all citizens to maintain social distancing, use hand sanitizers, and take effective precautionary measures to overcome this pandemic virus. The Indian government has made various initiatives and precautionary measures to safeguard the citizens from COVID-19 especially those who are residing in remote villages, as the awareness and knowledge about this deadly virus is found to be lesser in these areas [51].

In this pandemic condition, to offer health care services, the Government of India has taken up a decision for providing technology-based health care services for the rural people under common service centers. Various technologies-based health care services namely Teleconsultation services, Tele-Diagnosis, and general healthcare services have been developed and facilitated for the low-income people [1, 50]. In April 2020, the Indian government has designed "Arokiya Setu Application" for distancing the people from the deadly virus by spotting the regions where the COVID has been spread extensively. An application named "e-Sanjeevani" has been designed by the Government of India for providing teleconsultation services that associated people with the best health care professionals and medical practitioners through video conferencing [49]. Therefore, in the present world condition "health" has become a primary need and requirements for the people living in the society [12]. Hence this current research study has concentrated on determining the dimensions that are related with the rural citizens' trust for effectively utilizing the e-governance health care services. The research study has done an in-depth examination in analyzing the exact variable that increases the trust of rural citizens for adopting the health care services through ICT. Throughout the world there has been an enormous innovation and change has been found in governmental practices specifically in creating local governance [47]. In several ways the local governments have been linked with the higher level of government that includes the statutory, and constitutional frameworks, joint responsibilities for implementation of programmes, fiscal relations and in political perspectives. The definite series of linkages may vary significantly across the countries [11].

6.2   Determinants of Rural Citizens' Trust in e-governance Health Care Services During COVID-19

The term "Trust" is defined as the basic concern of an individual living in the country towards their priorities, differences and unified values. It is also determined as the citizens' expectations towards the government in facilitating services [21, 37]. Individual have their own system of beliefs towards government process, interactions and the activities with social and economic institutions [8]. People also rely on the behavior of the government officials and the political leaders in offering the services for satisfying their timely needs and requirements [22, 30]. In the developing countries, the vanishing state of trust among the poor people over government and their services are mainly due to the insufficient access and opportunities [9]. Intensification in accessing ICT by the government and the institutions has the extensive impacts in facilitating public services to the people through smartphones, social media, and websites [45].

Authors have stated that ICT is determined as a productive tool for gaining people trust on e-governance services through increasing effectiveness, transparency, cost effectiveness, and policy participation [35]. Offering e-governing services are found to be widely popular and the benefits acquired by the citizens are found to be remarkable. Government services through online are cheaper, faster, and easily accessible all around the world. Therefore the most of the developed countries are spending millions and lakhs of money for implementing and executing the digital projects [34]. Several researchers have examined the e-governance services is a platform to regain citizens' trust on functions of the government. Authors have highlighted that accessing internet tools like websites, digital payment systems, and social media helps in delivering the governmental services to the public. The productivity of these initiatives improves the interactions between the government and the citizens that ultimately increase the citizens' trust on government functions [44].

6.2.1 System Quality

The term "System" is defined as the association of different elements to work under a common objective (Hardcastle, 2011). System Quality is described as the efficiency of the Information System related to convenience, ease of use, reliability, and convenience, and various metrics & components. Intuitiveness, Sophistication, Flexibility, and Response time are the constructs that analyses the needs and values of the people [41]. The phrase "System Quality Ease of Use" is defined as the productivity of the e-governance system in accessing effectively by the people [20]. The variable "System Quality Functionality" is described as the efficiency of e-governance system related to its reliability and functionality for the people [31]. The study has concentrated on the elements related to the quality of e-governance services. The researchers have various dimensions that are associated with the quality of software to analyze the quality of e-governance system. High quality software increases the productivity and efficiency of the Information system [15].

6.2.2 Service Quality

An element "Service" is described as the association of inseparable, perishable, and intangible system efficiency that fulfills the user needs. TQM (Total Quality Management) Practices, balance card, ISO Six Sigma, and benchmarking are the various dimensions of service quality that examines the efficiency of e-governance services [5–7, 26]. Assurance, Empathy, Responsiveness, Tangibility are the major components of service quality in banking, tourism, and transport industries [39]. Behavior, Trust, Attitude, and Credibility are the preliminary factors that examine the service quality of the organization [25]. To analyze the satisfaction of the customers, the variables of the services quality such as reliability, responsiveness, personalization and quality are to be considered and measured [32]. The researchers have examined service quality model based on the operational issues with staff training and skills. The researchers have observed subjective, outcome, and process as the three major determinants for analyzing the service quality [23]. Authors have stated that the dimension service quality is called as the marketing strategy to improve the productivity and efficiency of the organization [3, 48].

6.2.3 Access to Information

The term "Access to Information" is defined as the freedom of expression of an individual that includes rights to seek as well as receive information from others. Access to Information is said to be a crucial need of an individual for receiving information fixed by law [33]. It is defined as the degree of transparency and accountability of sending and receiving the information between the citizen and government [18]. It is defined as the contemporary requirement for the students to access quality information through academic libraries [19]. Access to information is considered to be a most suitable platform for implementing good governance in the country [10]. In the developing countries, Information and Communication Technology (ICT) helps in handling several issues namely economic, social and health related problems [14]. It is applicable if both public and the government have access to information on decision making [17]. Rights to information access increases the public trust on government functions [13].

6.3 Research Gap

- Many studies have focused on the challenges and difficulties faced by the poor people in using e-governance services.
- Several authors have concentrated on explaining the theories and the dimensions associated to e-governance services.
- Several other researchers have done an empirical study on satisfaction of rural citizens on e-healthcare services. The Indian government has made multiple initiatives and spent a large amount of money in redesigning the mode of facilitating its services to the low-income group people through ICT. But still accessing those ICT based facilities are lacking behind in villages for seeking health care services through ICT among the rural people (Margaret et al., 2015).
- The present study is an attempt to analyze the level of rural citizens' trust on accessing e-governance health care applications during this epidemic condition of COVID-19.
- The present research study has examined the hidden layers that exist between the variables used in the study by adopting neural network analysis through R studio. Through this analysis, the research study has analyzed the exact dimension that is strongly associated to the output variable "Rural Citizens' Trust" on accessing e-governance health care applications

6.4 Theoretical Framework

The present study has adopted DeLone and McLean [16], Information Success Model for constructing the theoretical model [28]. In this model the authors have analyzed

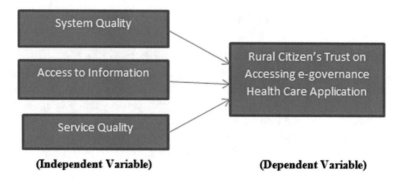

Fig. 6.1 Rural citizens' trust model for e-governance health care application usage. *Source* Authors' Model

the net benefits obtained by the users through accessing the specific system. Figure 6.1 highlights the conceptual model framed for the research study. The authors have considered "System Quality", "Access to Information", and "Service Quality" as the three primary factors for measuring the outcome variable "Rural Citizens' Trust" for accessing e-governance health care applications.

6.5   Research Methods

The researchers have designed the questionnaire for examining the Rural Citizens' Trust on accessing e-governance health care applications. Descriptive research design has been adopted and purposive sampling technique has been utilized by the authors for conducting the research study depending on the specific phenomenon. The authors have considered Kanchipuram district, as its rural literacy rate is seems to be 76% that is lesser than the average literacy rate of Tamil Nadu (80.23%). Moreover, the Common Service Centres (CSCs) of Kanchipuram district is found to be 41 for the entire population size of 3,998,252. This highlights that for the population size of 97,518 only one CSC can be assessable by the rural people. Therefore, the authors of the present research study have considered Kanchipuram district for investigation. Through the random sampling technique, the questionnaire has been distributed to the 500 rural respondents of randomly selected villages for gathering the primary data. The demographic statistics of the respondents from rural villages has been displayed in Table 6.1. Only 10.2% of rural respondents are accessing the e-governance health care applications. From the neural network analysis, the specific dimension that are strongly associated with the outcome variable "Rural Citizens' Trust" have been identified by the researchers.

Table 6.1 Rural respondents' demographic statistics

Variable	Description	Frequency	Percentage
Gender	Male	255	51.3
	Female	245	49.7
Age (Years)	18–25	150	30.5
	26–35	151	30.4
	36–45	112	22.2
	46–55	60	12
	Above 55	27	5.3
Marital status	Married	460	91.2
	Unmarried	40	8.1
Occupation	Farmer	267	53.4
	Job	139	28.2
	Own Business	73	15.3
	Land Labors	21	3.8
Education	Below 10th Std	125	25.6
	10th Std	111	22.2
	12th Std	144	28.8
	Graduate	120	23.4
Income	Rs.20000–50000	188	37.2
	Rs.50001–1Lakh	278	55.6
	Greater than Rs.1Lakh	36	7.4
Frequency of accessing e-governance health care applications	Never	69	13.8
	Rarely	147	29.4
	Sometimes	180	36.2
	Often	54	10.4
	Always	50	10.2

6.6   Data Analysis

For deriving the statistical analysis SPSS (Statistical Package for Social Sciences) has been used. Through pilot study the reliability of the questionnaire framed by the researcher has been measured. From the Kanchipuram district 500 rural respondents have been selected and distributed the questionnaire. From the reliability test, the cronbach's alpha value of 0.87 has been generated. Therefore, the internal consistency of the questionnaire is found to be reliable [43]. Through neural network analysis, the current research study has trained the gathered data from the rural respondents and has examined the exact determinant that is highly associated with the output variable.

6.6.1 Machine Learning Technique

The present research study has taken considered Neural Network analysis as machine learning technique for examining the specific dimension that are related strongly with the outcome variable. Through the structured algorithm of neural network algorithm of machine learning technique in r studio, the neural network analysis has been made by the researchers. The term "neural network" is determined as the neuron that is highly related with the neuronal system of our human body. Each neuron in our human body is associated with the dendrites that passes the signals to another neurons in the shape of electrical impulses and these impulses are received by our human body to make particular actions [36]. The input has been received as an activation code and through which a specific action is made by the humans [29]. By obtaining multiple inputs, the human system performs the action through strong examination and analysis for taking an appropriate and correct decision [27]. A specific weight called 'w' has been measured for each input variable used in the study. Through the summary function, the weights of input variable have been calculated through examining true and false judgements. The activation of summary function is done through limit(). The process has been continuously tested by the binary, numerical and categorical variables and the node called "Single Level Perception" is measured. Output, Hidden, and Output are the three different categories of Neural Networks. The model formulation of neural network analysis for the present research study is highlighted in Table 6.2.

The current study has 500 data sample with "Service Quality", "System Quality", and "Access to Information" are the three different dimensions taken by the researcher to measure the outcome variable "Rural Citizens' Trust". Figure 6.2 has highlighted that the determinant "Access to Information" is found to be a highly associated with the output variable "Rural Citizens' Trust" with the highest score "3.29325". This shows that the output variable "Rural Citizens' Trust" has acquired highest score of "5.23573" by extracting the maximum weights from the dimension "Access to Information".

Table 6.2 Model formulation

S. No	Name of the layer	Subscript	Description
1	Input layer	$x(1,2,3,4...m)$	Input variables used in the study
2	Hidden layer	$h(1,2,3...n)$	Hidden weights extracted from the input layers
3	Output layer	y	Output variable analyzed from the weights of hidden layers

Source Lim (2018)

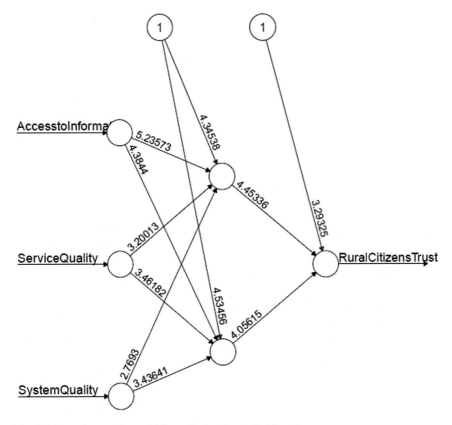

Fig. 6.2   Neural networks model for analyzing the rural citizens' trust

6.6.2   Program Code of Machine Learning Technique Neural Network Analysis for Predicting the Rural Citizens' Trust

```
setwd("C:/Users/Bhuvi/Desktop")
getwd()
egdata = read.csv("gov.csv", header = TRUE)
library("neuralnet")
#install.packages("neuralnet")
require(neuralnet)
set.seed(444)
```

```
n <- neuralnet(RuralCitizensTrust ~ AccesstoInformation + ServiceQuality + SystemQuality,

    data = egdata,

    hidden = 2,

    linear.output = FALSE)
plot(n)
```

6.7   Findings and Discussion

The demographic statistics of the respondents is displayed in Table 6.1. This shows that only 10.2% of the respondents from villages of Kanchipuram district are accessing e-governance health care applications. The present research study pointed out the rural citizens' needs and requirements for using e-governance health care applications that are to be fulfilled for the low-income group people. Various findings of the research study have highlighted those biometric issues, crowding, electricity and internet issues, closure of CSCs (Common Service Centres) as the obstacles faced by the poor people residing at villages for productively accessing e-governance services [5–7].

Several researchers have examined that offering health care services especially for the group of the aging population residing in remote villages are found to be highly challenging. Accessing e-health care applications is the only solution to overcome the challenges faced by both medical practitioners as well the rural people. Many researchers have analyzed that facilitating health care services for the elderly population living in villages is seems to be challenging. Lack of infrastructure and awareness are the major problems related with the system of e-governance health care application [34]. In many developing and developed countries, several productive and innovative applications has been designed and developed and it has been productively accessed by many people. But, when the remote villages are considered, the knowledge, awareness and level of frequency of accessing e-health care applications are seems to be inadequate. All the clinics and hospitals at remote villages are intensified with the provision of quality computers, internet, and with all supported accessories for the potential execution of e-governance health care applications [40]. Uneven power supply is the challenge encountered by the rural citizens for frequently accessing e-governance health care services [40].

The current research study has examined that the determinant "Access to Information" is the primary factor towards rural citizens' trust on accessing e-governance health care applications. The authors have designed the theoretical model and it has been tested by using neural network analysis in r studio. Through the three input layers namely "Service Quality", "System Quality", and "Access to Information" the study has identified that the dimension "Access to Information" is found to be strongly associated with the outcome variable "Rural Citizens' Trust". The findings of the research study convey that "Access to Information" is said to be a crucial need of an individual for receiving information fixed by law [33]. Therefore, the findings of the research study highlight that before developing or designing any application related to health care services for the low-income group people, the medical practitioners, government officials, and the application developers should focus on rural peoples' primary need and requirements to regain their level of trust on accessing e-governance health care applications.

6.8 Conclusion

Initiatives under e-governance health care services are said to be developing field particularly in the current epidemic situation of COVID-19. E-governance health care applications are called as the conversion of medical informatics and public health through facilitating ICT based health care services. In India, it is found to be highly challenging to deliver e-governance health care services for the rural people. The present study has made an utmost contribution in analyzing the dimension through neural network analysis that is strongly related with the rural citizens' trust on accessing e-governance health care applications.

References

1. Achampong, E. K. (2012). *Electronic Health Record System: A Survey in Ghanaian Hospitals*, Vol. 1 (p. 164).
2. Armstrong, L., & Gandhi, N. (2012). Factors influencing the use of information and communication technology (ICT) tools by the rural famers in Ratnagiri District of Maharashtra, India. In *Proceedings of The Third National Conference on Agro-Informatics and Precision Agriculture 2012 (APIA 2012)* (pp. 58–63).
3. Asubonteng, P., McCleary, K. J., & Swan, J. E. (1996). SERVQUAL revisited: A critical review of service quality. *Journal of Services Marketing, 10*(6), 62–81.
4. Bala, M., & Verma, D. (2018). Governance to good governance through e-governance—A critical review of concept, model, initiatives & challenges in India. *International Journal of Management, IT and Engineering, 8*(10), 224–269.
5. Bhuvana, M., & Vasantha, S. (2020). Assessment of rural citizens satisfaction on the service quality of common service centers (CSCs) of e-governance. *Journal of Critical Reviews, 7*(5), 302–305.
6. Bhuvana, M., & Vasantha, S. (2020). Determinants of behavioral intention to access e-governance services by rural people with the mediating effect of information and communication (ICT) literacy. *Journal of Advance Research in Dynamical and Control Systems, 12*(2), 176–187.
7. Bhuvana, M., & Vasantha, S. (2020). Neural network machine learning techniques using R Studio for predicting the attitude of rural people for accessing e-governance services. *Journal of Advanced Research in Dynamical and Control Systems, 12*(2), 2552–2562.
8. Chadwick, A., & May, C. (2003). Interaction between states and citizens in the age of the internet: "e-Government" in the United States, Britain, and the European Union. *Governance, 16*, 271–300.

9. Cheema, G., & Popovski, V. (2010). *Engaging Civil Society: Emerging Trends in Democratic Governance.*
10. Cletus, D. (2011). Access-to-information legislation as a means to achieve transparency in Ghanaian governance: lessons from the Jamaican experience. *IFLA Puertorico.*
11. Cortes, E. Jr. (1997). Contribution to the panel on building civil society. In R. H. Wilson, & R. Cramer (Eds.), *International Workshop on Governance: Third Annual Proceedings* (pp. 19–22). University of Texas, Lyndon B. Johnson School of Public Affairs.
12. Currie, et al. (2015). *BMC Health Services Research*, Vol. 15 (p. 162).
13. Darch, C., & Underwood, P. (2010). *Freedom of information and the developing world. The Citizen, the State, and Models of Openness.* Chandos Publishing.
14. Davis, F. D., Bagozzi, R. P., & Warshaw, P. R. (1989). User acceptance of computer technology: A comparison of two, management science. *Management Science., 35*(8), 982–1003.
15. Dawes, S. S. (2008). The evolution and continuing challenges of egovernance. *Public Administration Review, 68*(s1), S86–S102.
16. Delone, W. H., & McLean, E. R. (2003). The DeLone and McLean model of information systems success: A ten-year update. *Journal of Management Information Systems, 19*(4), 9–30.
17. Delponte, L., Grigolini, M., Moroni, A., Vignetti, S., Claps, M., & Giguashvili, N. (2015). ICT in the developing world. *Scientific Foresight Unit.* https://doi.org/10.2861/52304
18. Easton, D. (1965). *A Systems Analysis of Political Life.* John Wiley.
19. Fang, C. (2013). *Chinese Librarianship in the Digital Era.* Science Direct.
20. Farazmand, A. (Ed.). (2018). *Global Encyclopedia of Public Administration, Public Policy, and Governance.* Springer.
21. Fishbein, M., & Ajzen, I. (1975). *Belief, Attituak, Intention and Behavior.* Addison-Wesley.
22. Fishbein, M., Ajzen, I., & Hinkle, R. (1980). Predicting and understanding voting in American elections: Effects of external variables. In I. Ajzen, & M. Fishbein (Eds.), *Understanding Attitudes and Predicting Social Behavior.* Prentice-Hall.
23. Galloway, L., & Ho, S. K. (1996). A model of service quality for training. *Training for Quality, 4,* 20–26.
24. Government of India, India in Figures 2015 (Ministry of Statistics and Program Implementation, New Delhi: Central Statistics Office, 2015).
25. Gronroos, C. (1984). A service quality model and its marketing implications. *European Journal of Marketing, 18,* 36–44.
26. Gupta, K. P., Singh, S., & Bhaskar, P. (2016). Citizen adoption of e-government: A literature review and conceptual framework. *Electronic Government, An International Journal, 12*(2), 160–185.
27. Gurney, K. (2004). *An introduction to Neural Network.* UCL Press Limited is an imprint of the Taylor & Francis Group.
28. Jafari, S. M., Ali, N. A., Sambasivan, M., & Said, M. F. (2011). A respecification and extension of DeLone and McLean model of IS success in the citizen-centric e-governance. In *2011 IEEE International Conference on Information Reuse and Integration*, Las Vegas, NV, (pp. 342–346).
29. Johnson, S. A. (2004)0. *Neural Coding Strategies and Mechanisms of Competition.* Cognitive Systems Research.
30. Jon Welty Peachey Laura J. Burton Janelle E. Wells. (2014). Examining the influence of transformational leadership, organizational commitment, job embeddedness, and job search behaviors on turnover intentions in intercollegiate athletics. *Leadership and Organization Development Journal, 35*(8), 740–755.
31. Kolsaker, A., Lee-Kelley, L. (2007). 'Mind the Gap II': E-Government and E-Governance. In: M.A. Wimmer, J. Scholl, Å. Grönlund (Eds.), *Electronic Government. EGOV 2007.* Lecture Notes in Computer Science, vol. 4656. Springer. https://doi.org/10.1007/978-3-540-74444-3_4
32. Lee, G., & Lin, H. (2005). Customer perceptions of e-service quality in online shopping. *International Journal of Retail & Distribution Management, 33*(2), 161–176.
33. Lipinski, T. A., & Britz, J. (2000). Rethinking the ownership of information in the 21st century: Ethical implications. *Ethics and Information Technology, 2,* 49–71.

34. Mahmood, M. (2019). *Does Digital Transformation of Government Lead to Enhanced Citizens' Trust and Confidence in Government?* Springer.
35. Moon, M. J. (2003). Can IT help government to restore public trust?: Declining public trust and potential prospects of IT in the public sector. In *Proceedings of the 36th Annual Hawaii International Conference on System Sciences (HICCS 2003), Hawaii, HI, USA* (pp. 1–8).
36. Noon, H. (2013). *Artificial Neural Network: Beginning of the AI revolution.*
37. Nooteboom, B. (2010). The dynamics of trust: Communication, action and third parties. *Revista de Economía Institucional, 12*(23), 111–131.
38. Nzaramyimana, L., et al. (2019). *Procedia Computer Science, 161*, 350–358.
39. Parasuraman, A., Zeithaml, V. A., & Berry, L. L. (1988). SERVQUAL: A multiple-item scale for measuring consumer perceptions of service quality. *Journal of Retailing, 64*(1), 12–40.
40. Rudolph, et al. (2014). A study of the issues of eHealth care in developing countries: The case of Ghana. In *Twentieth Americas Conference on Information Systems, Savannah.*
41. Saxena, K. B. C. (2005). Towards excellence in e-governance. *International Journal of Public Sector Management, 18*(6), 498–513.
42. Singh, M., & Sahu, G. P. (2018). Study of e-governance implementation: A literature review using classification approach. *International Journal of Electronic Governance, 10*(3), 237–260.
43. Tavakol, M., & Dennick, R. (2011). Making sense of cronbach's alpha. *International Journal of Medical Education, 2*, 53–55.
44. Ud Doullah, S., & Uddin, N. (2020). Public trust building through electronic governance: An analysis on electronic services in Bangladesh. *Technium Social Sciences Journal, Technium Science, 7*(1), 28–35.
45. United Nations. (2018). *United Nations e-government survey 2018: Gearing e-government to support transformation towards sustainable and resilient societies.* United Nations publication.
46. Varun Kumar, M., & Venugopal, P. (2015). E-governance and rural development (A study specially focused on villages of KatpadiTaluk, Vellore District Of Tamil Nadu). *International Journal of Applied Engineering Research, 10*(92), 319–324.
47. Wilson, R. H. (2000). Understanding Local Governance: an international perspective. *Revista de Administração de Empresas, 40*(2), 51–63.
48. Zeithaml, V. A., Berry, L. L., & Parasuraman, A. *Journal of Marketing, 60*, 31–46.

Website Sources

49. Centre for Development of Advanced Computing. (2020). *E-Governance.* https://www.cdac.in/index.aspx?id=st_egov_eSangam
50. CSC Health. (2020). *E-Health Care Services.* http://cschealth.in/Portalclass/index/services
51. United Nations Member State. (2018). Retrieved 18 June 2020 from https://visit.un.org/sites/visit.un.org/.../FS_List_member_states_Feb_2013.p
52. World Health Organization (WHO) Publications. (2018). *Rural Population.* https://www.who.int/publications

Chapter 7
Artificial Intelligence (AI) Use for e-Governance in Agriculture: Exploring the Bioeconomy Landscape

Dimitris C. Gkikas, Prokopis K. Theodoridis, and Marios C. Gkikas

Abstract This chapter examines how artificial intelligence (AI) applications can leverage agriculture through electronic government (e-government) public services and big data. The constant use of public services daily generates enormous volumes of data. Fueled with these data, AI is introduced as a disruptive innovative mechanism, which is expected to play a positive role in making governments work effectively, transforming the way that public services are delivered. However, the positive impact of "smart" technologies should not be taken for granted. Technologically advanced countries invest on technology expansion and build more service-oriented, competent, and transparent government digital public services aiming to improve citizens life quality. A part of digital public services refers to bioeconomy where agriculture cannot afford to be absent from these technological innovations. Ministries of Agriculture implement e-government structures to promote agricultural development, improving farmers life quality, promoting sustainable agriculture management and decision-making abilities, and encouraging stakeholders to participate. This chapter highlights the AI technology advancements, exposing the information models, and exploring the influence of factors on digital public services implementation in bioeconomy. It identifies the stakeholders of a mutual collaboration between the citizens and the state, providing segmented and structured information about the benefits of AI appliance to agricultural economy, and the role of big data in digital public services. Technologically advanced countries are also examined for AI initiatives, investments, and appliances on e-government agricultural services. Finally, a future AI e-government system assessment framework is introduced.

D. C. Gkikas (✉)
Athens University of Economics and Business, 76, Patission Str., GR-10434 Athens, Greece
e-mail: dgkikas@aueb.gr

M. C. Gkikas
University of Patras, 1 M. Alexandrou Str, 26334 Koukouli, Patras, Greece
e-mail: mario.gkikas@upatras.gr

P. K. Theodoridis
Hellenic Open University, 18 Aristotelous Str., 26335 Patras, Greece
e-mail: proth@eap.gr

© The Author(s), under exclusive license to Springer Nature Switzerland AG 2023   141
C. Gaie and M. Mehta (eds.), *Recent Advances in Data and Algorithms for e-Government*,
Artificial Intelligence-Enhanced Software and Systems Engineering 5,
https://doi.org/10.1007/978-3-031-22408-9_7

Keywords e-government services · Digital governance · Artificial intelligence ·
Smart farming · AI bioeconomy · Digital agriculture

7.1 Introduction

The world's population will reach 10 billion by 2050. AI adoption seems capable
to drastically bring a solution to the upcoming lack of food supply. This could be
solved through the agriculture value chain making agriculture be more efficient,
productive, and sustained. Scientific research and applications in smart agriculture
remains scarce. E-government refers to government's digital services which improve
public services quality and eliminate the distance between the state and the society
in order to reduce the cost and increase the interoperability among e-government
sectors. Governments should invest on AI-based services in agriculture to solve problems like the farm supply, production, and restoration, quality testing, stockpiling,
trading, distribution and selling [23].

AI increasingly dominates the hardware and software platforms and the internet
of things services (IoT). Huge amounts of investments in AI have led to a point
where this technology exists in people's transactions including navigation, voice-
recognition, smart phones, finance, logistics, trading, language detection, smart
homes, image-recognition, augmented reality, robotics, sentiment analysis etc. The
global AI market revenues were estimated at 120 billion USD in 2022 and it is
expected to exceed 1597 billion USD by the year of 2030 (Fig. 7.1).

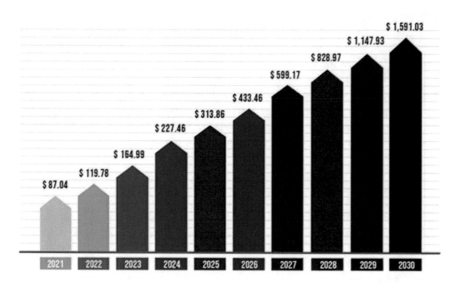

Fig. 7.1 Market size from 2021 until 2030 (USD Billion) (Reproduced from Precedence Research
2021)

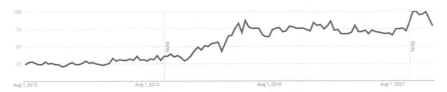

Fig. 7.2 Artificial intelligence interest from 2012 until 2022 (Reproduced from Google Trends [25])

The estimated value of AI interest in 2022 is between 79 and 99 points worldwide (Fig. 7.2) [25].

The annual estimated value of AI is between 3.5 and 5.8 trillion USD across the industries, and it is estimated that 70 percent of businesses will use AI by the year of 2030 (Fig. 7.3) [9, 56].

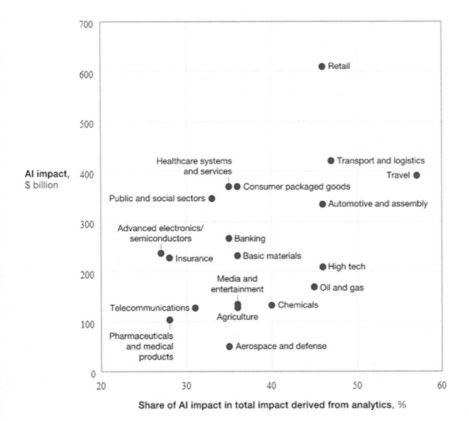

Fig. 7.3 Artificial intelligence has the potential to create value across sectors (Reproduced from Chui et al. [9])

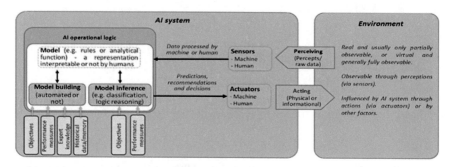

Fig. 7.4 AI system (Reproduced from OECD [45])

7.1.1 Artificial Intelligence (AI)

According to the "A guide to using artificial intelligence in the public sector" report from the Office for AI service of the Government Digital Service (GDS) in United Kingdom (UK), "*AI is a research field spanning philosophy, logic, statistics, computer science, mathematics, neuroscience, linguistics, cognitive psychology, and economics. AI can be defined as the use of digital technology to create systems capable of performing tasks commonly thought to require intelligence*". There is not a universally accepted term of AI and there are government agencies, organizations, researchers, communities, and other scientific entities which have provided a variety of AI definitions [24].

An AI system structure consists of three entities: sensors, operational logic, and actuators. Sensors are used to identify and collect raw data from the environment.

Given a set of objectives, the operational logic handles the input data from the sensors managing an output for the actuators (humans or machines). The output refers to predictions, suggestions, or decisions from which the actuators react to change the state of the environment (Fig. 7.4) [45, 58].

7.1.2 AI in the Public Sector

Many governments around the globe are researching how to utilize AI for decision-making in the public sector. The world's wealthiest and technologically advanced countries compete to develop and implement artificial intelligence applications that could define citizens lives. AI advances provide great benefits to social well-being in areas such as medicine, bioeconomy including agriculture and aquaculture, environment, and public welfare [20].

Among the strongest financially countries, United States of America (USA) competes other countries to keep the lead in AI innovations. China has committed to invest 150 billion USD in research and AI tech companies aiming for global AI leadership by 2030. The European Union (EU) works more methodologically, trying to take the lead in setting the rules for a constitutional AI regulation which will manage to control the boundaries or AI usage [56].

The European Commission published a white paper dedicating a part for AI promotion the adoption by the public sector [14].

Based on the Oxford Insights research, there are evidence about 160 countries readiness for using AI in public sector. Based on the scientists' findings the top three countries for AI adoption are the USA, Singapore, and UK. There is 40% of countries which has published draft editions of AI strategies where 30% of the countries have published their AI strategies and 10% are drafting one, showing that AI rapidly emerges. National AI strategies around the world vary based on the geographic location, government, technology sector, data, and infrastructure. The northern countries seem to progressively adapt to the new AI era way better than the southern countries do (Fig. 7.5) [53].

Regarding the countries national AI strategies, their goal is to gradually regulate the public sector administrations and help them operate within highly abstract and complicated political, cultural, and socio-economic ecosystems including actors like intermediaries, NGOs, and the civil society, operating at the same time on a multi-level structure including international, national, regional, and local level. Thus, any AI dedicated service must be efficient to provide solutions and answer complex demands with different goals and objectives in multilevel environments. Governments regulate the activity of citizens, the use of public power, and maintain the public order. Designing and activating powerful services through AI technology require highly sophisticated control protocols based on the public administrative law, preserving the ethical and human rights principles (Leslie et al., 2021).

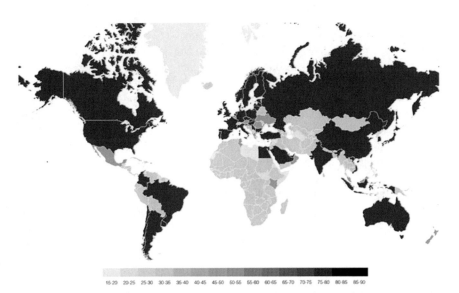

Fig. 7.5 National AI strategy (Reproduced from Oxford Insights 2021)

Public administrations' goals are to protect their citizens and provide community services promoting the citizens' well-being. Despite that the biggest fraction of AI technology is held by industry and academic research, there are also AI projects which apply to the public sector. Progressed governments already make use of AI applications in many fields. Due to the goals of the governments for civil protection, AI adoption by the governments appears to be trivial. They must incorporate regulations, public policies, and laws for the society and the best of the citizens and their privacy on one hand and follow the technological advancements of the private sector on the other hand [35, 52].

The National Association of State Chief Information Officers (CIOs) and the Public Technology Institute in USA published the results of a conducted research showing the impact of using IT automations on government agencies. The survey revealed that the 67% of the state CIOs believe that AI and machine learning (ML) will be the most important disruptive technologies over the next five years. Almost 72% of CIOs across the USA are already planning, implementing, or using AI applications. Through AI, CIOs aim to remotely monitor, manage, and control chronic diseases, sustainable agriculture, cybersecurity, waste, fraud, abuse, optimize traffic constraints, strengthen the public trust to the new technologies and promote the necessity for human centered services [37, 50].

While most governments agencies try providing solutions in more basic public administration problems it would be beneficial for them if they would prepare for the forthcoming new technology by making investments on a steady regular pace in order to be on track with the modern operating standards. Taking into consideration the best practices of other more technologically advanced governments and industries, they could design their next AI government strategy for public administration [37].

There are six strategies that can help governments design their AI public sector strategy: (1) Make AI a part of a government goals, (2) Design and implement a citizen-centric program, (3) Ask from citizens to engage with this program, fourthly is to build on the existing resources, (5) Be data-prepared, data-sensitive, data-concerned and very careful with the data privacy, (6) Mitigate ethical issues avoiding AI decision making; and, train employees instead of replacing them (Fig. 7.6) [37].

Although governments around the world are in a rush to adopt e-government services. Especially the developing countries still have to overcome obstacles associated to multiple contextual factors including insufficient financial and technological resource limitations, and they lack digital adoption and infrastructures, data security, and cybersecurity. Apart from ICT infrastructure, e-governance transformation and adaptation also faces another challenge. The physical and digital environment produces information which needs to be managed and in order to be properly used for the digital governance. An information model called "The Onion Ring" was introduced. The Onion-Ring model maps the environment from which the information is generated (Fig. 7.7) [28].

In order this to happen, governments should invest on data acquisition. The United Nations E-Government Development Index (EGDI) measures the e-government generated data which are responsible for the e-government development from countries around the world. E-government data are daily generated through constant

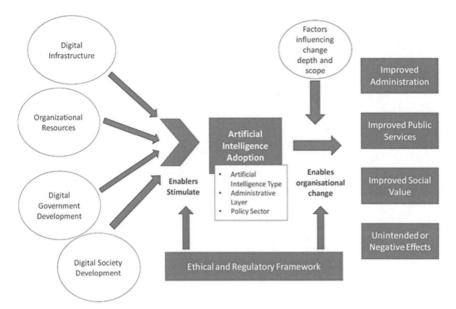

Fig. 7.6 Conceptual e-government system (portal) using AI technology (Reproduced from Misuraca and van Noordt [39])

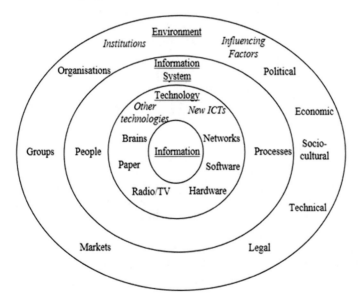

Fig. 7.7 The onion-ring model (Reproduced from Heeks [28])

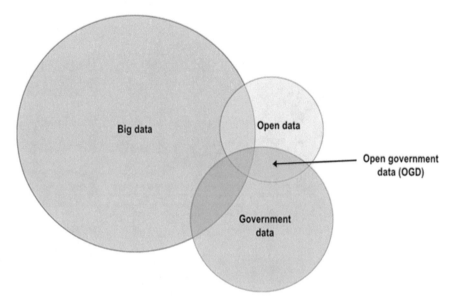

Fig. 7.8 Big data relationships in the public domain (Reproduced from United Nations [69])

electronic transactions among people and government digital services. According to United Nations survey, there are several types of big data and various terms are used to describe them including types such as the public, government, census and survey, administrative, open government, big, geospatial, and real-time data. Most of these types have been used in different occasions. The context of each data type refers to different use and purpose (Fig. 7.8) [69].

7.1.3   AI in Bioeconomy

According to the European Commission, the bioeconomy combines the specific parts of the economy that use renewable natural planet resources to produce food, biomaterials, and energy. Bioeconomy lies on the sustainable production and transformation biomass into of bioproducts, biomaterial, and energy. There are numerous strategies which indicate the roadmap of the transition to bioeconomy [16].

Artificial Intelligence is a foundational technological advancement which provides numerous and multilevel advantages to circular bioeconomy, including environmental sustainability, public healthcare, transportation, education, public safety, cybersecurity, economic and bioeconomic growth [65].

AI have the potential to provide an enhanced and sustainable biomass production within agriculture, aquaculture, and forestry. Biomass refers to the primary sector biological resources including food, chemicals, oil, and gas [63].

Regarding the biomass supply AI improves yield, it decreases the fertilization quantities, it makes proper use of carbon and water. Machine learning and data mining techniques are used to predict environmental impact and soil conditions on the fields [16].

Being able to lead the new digital era by gathering and analyzing the bioeconomic data in a sophisticated and complicated way, AI is able provide the machines with the ability to collect, process and analyze data aiming to make them gradually capable to learn from the entire process [33].

Aiming to increase the performance and the productivity in a sustainable circular way, big volumes of data are gathered and processed. Sensors, robots, farm fishing and forestry machines, are able to collect, store and analyze data. Automated robots handle the agriculture important tasks of fertilizing, planting, and harvesting. Machine learning and deep learning algorithms are also used to analyze data collected from sensors like cameras, satellites, drones, and airplanes collect data from the farms, air, sea, and fires [33].

Due to the processing of millions of recorded data instances the prediction accuracy is optimized, and the systems acquire the ability to learn for themselves making the decision-making process more efficient. The huge volume of gathered data including data from weather forecasts is analyzed, processed, and transformed into sophisticated patterns which can make prediction for supply sustainable biomass at a decreased cost also generating an added value to the entire process with benefits for the environment, the farmers, the investors, and the consumers [33].

These scientific approaches are considered the tools which will help with the development of the European and the world bioeconomy to proceed in a more sustainable development [16].

An IoT generic framework for agriculture (Internet of Farming, IoF) consists of six layers (Fig. 7.9) [3]:

- The physical layer
- Network Layer
- Middleware Layer
- Service Layer
- Analytics Layer
- User Experience Layer

7.2 Literature Review

In today's global government systems, the disruptive technologies of artificial intelligence, big data, IoT, decision making, and cloud services have established themselves as the main particles of restoration of the global communication and economy. Governments need to be on the edge of technology to understand what society's demands are and activate the government actors to conduct data analysis and decision making using raw data from multiple resources. The main axes of the new age

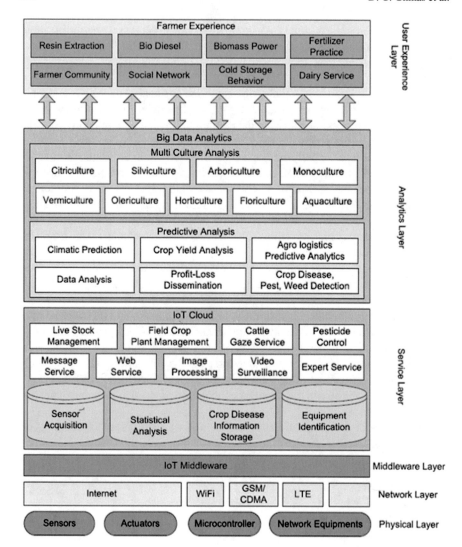

Fig. 7.9 Generic IoF application model in bioeconomy (Reproduced from Bansal et al. [3])

smart governance include some of the most important services and processes of a state like applications of AI in education, data extraction techniques, digital democracy, geopolitics, supply chain systems, taxation, cybersecurity, and fraud prevention [59].

AI applications could bring benefits to the public sector including information processing for data collection from multiple source and interpretation into knowledge. AI could change the citizens perspective of the environment. Decision making processes would activate the autonomous action taking for processes which could be completed by machines. AI could easily lead to completion certain goals and objectives. Enhanced technologies increase the potential of more opportunities for the public sector including the public administration and decision-making performance shaping the relationship between the citizens and the government by providing more citizen centric high-quality services, increasing the citizens engagement in the activities of the public sector [36].

Almost fifty countries around the world have designed or implemented national AI strategies or policies to set boundaries on how AI can be applied and operate on government services and processes. A strategic planning of goals and objectives on how they can be strategically achieved aims to help countries build a roadmap of AI milestones, fully aligned to the structures, regulations and culture of each country actors and government ecosystem. While this implies that a significant majority of countries are not yet planning a strategy, it does indicate that many countries now see AI as a national priority. Most of the countries have or design an AI policy for economic growth. By the end of 2020, the EU had increased research and development investments in AI by 20 million Euros and 1.5 billion Euros in research funding [5].

The AI policy which will be used depends on the government sector of application. Due to the nature of some sectors to daily generate data on the field, there are certain government sectors which are considered to be more suitable than others for AI implementation. There is significant lack of AI adoption in sectors such as bioeconomy and agriculture. Analyzing the different policy required in bioeconomy including agriculture, aquaculture, and forestry in which the AI systems can be extremely efficient governments could track which factors are related to the adoption of AI and how it generates a series of advantages. AI technologies are expected to enhance the development of smart farming and agriculture and improve the entire food and production supply chain [38].

There is an increasing demand in agriculture for protection from the unstable weather, food safety, and big production in low cost. AI systems true contribution is when they are used for data observation, pattern creation and decision-making enhancement. AI implementation in agriculture drastically respond in diseases by providing highly accurate solutions up to 98%. In order this to happen big data collection is essential. Farmers are called to respond to the new demands of collecting and processing the data that are generated from the production procedure and they will support the overall production quality. Among others, there is a strong need to secure big data collection from various sources. Data collection, storage, visualisation, and security are essential and require producers' attention. More accurate data will manage to provide better production, bigger profits, and smaller environmental footprint [7].

Transforming traditional farming into smart farming or precision agriculture is essential to proceed to the next generation of massive food scale production. Currently, smart farming adopts advanced AI technology for agriculture activities. AI has altered entire agriculture sectors in the context of business modeling, designing and implementation analysis of blue river technology project. AI impact in the new age agriculture includes precision farming, machine and deep learning patterns, internet of farming (IoF), software as a service (SaaS), and advanced forecasting [54].

Due to economic recession, war, and climate change the global food production faces multiple threats and dangers. The agriculture food business is in the middle of a production and food management instability. Recently, the AI adoption has provided farmers the capability to increase the average production, decreasing costs, preserve natural resources, minimize the environmental footprint, decreasing human escalation and all these in a sustainable way. AI pillars of learning, observing, analyzing, modeling, and problem solving contributes to the agriculture food (agri-food) industries food management crisis. AI systems adoption in agriculture contributes towards the efficiency improvement when it comes to quality standards of crop yield, soil condition, careful use and preservation of water supply, fertilizations, pesticides, etc. [55].

Globalization has generated new needs and processes in the way markets operate in global scale. E-commerce growth has altered the rules of trading and dispatching. Agriculture (agri) business value chain could play an essential role in the global agri-business sector increasing the agriculture profits, improving competitiveness, lowering the costs, and improving the food quality. However, there are evidence which show that there are certain value chain stages which are underdeveloped or underestimated in comparison to other more developed stages. The processing and customer services sectors seem to be lower in the list of how agri-food businesses perceive management excellence. Despite the drawbacks, AI agri-food business value chain manages to expand its logic in different sectors such as water management, weather prediction, yield, cost, supply, demand, energy, crop planning, consultation, consumer behavior etc. [23].

7.3   Research Methodology

The current chapter is based on a literature review regarding the application of AI in e-government systems in bioeconomy by public administrations across the countries. The purpose of this research is to map the stages of AI adoption and appliances among the most technologically advances countries, highlighting the progress of e-government services in digital agriculture advanced countries, revealing the potential of AI models in e-government systems, proposing a framework assessment for future AI e-government services evaluation.

The authors have conducted a research based on scientific terms and synonyms including e-government system, e-government frameworks, AI in agriculture, AI in e-government services, AI in bioeconomy, internet of farming, internet of things in agriculture, smart farming, digital agriculture, e-government policy, AI frameworks for e-government etc. International repositories and databases have been used along with international organizations and official state departments and agencies files, reports, and projects.

Among the most technologically advanced countries including USA, China, UK, Israel, India, Canada, Singapore, Japan, and Korea and the EU which are investigated for their progress in AI applications in e-government services for agriculture promotion. The research includes annual national investments in AI, national AI strategies, national AI initiatives and e-government services in AI.

The main goal is to provide an overview of an AI e-government portal assessment framework approach for agricultural services across the countries focused on examining whether an e-government system serves the public good and increases the farmers satisfaction and experience.

A future version of this chapter will provide the architecture of an assessment framework for AI e-government system. This framework will record and assess the e-government agriculture portals in terms of provision of digital services based on set of identified e-services from the best practices of the most technologically advanced ranking countries e-government portals which use AI. In order to assess the usability and overall contribution of AI application in an agricultural government information system (portal), this framework will measure the maturity and service efficiency based on how governments manage to provide a wide spectrum of e-services to its people. It will provide statistical data on what level the government administration responds to the farmers transactions and needs by quantifying factors like services, monitor services, recommendation services, weather forecast services, search information, education services, download information, download documents, data privacy and security, website security, disability access, foreign language support, advertisement and online payment system, business promotion, government advertisement, expositions etc., consist a portion of the total number of measured indicators provided by a government. Through a scoreboard, this framework will measure the range of e-government digital services that make use of AI technology indicating the actual level of digital readiness or any lack of basic structures indicating its class (A, B, C) [34].

7.4 AI Models in Agriculture

Agricultural innovation companies are implementing AI projects to model maps and predict potentially correlated factors which can have beneficiary results to the food production using geographic information systems (GIS) technology. By monitoring and collecting data using sensors, satellites, and drones from huge data resources like weather, temperature, images, water consumption, ocean temperature, deforestation, fires, soil quality, animals feeding etc. AI systems are able to create patterns which could improve the farms production with highly detailed precision. AI monitoring systems help to dynamically use the earth's natural resources avoiding any water, or food spoilage [71].

A sophisticated AI system can gather the data and generate patterns knowing when the growing season begins, it assists with planting, growing, and harvesting seasons. Image recognition and computer vision systems detect any possible plant pathogens, pests, or diseases, and they monitor the possible harvesting dates in order to reduce any product, food, or water waste. In places where the water is in shortage or the weather is unstable like in Africa, AI systems provide recommendations to farmers about the potential date during the season that it is possible to rain and warn farmers to be ready for planting or harvesting aiming to eliminate the risk of crops, animals, or production loss [71].

Smart technologies collect and model data for farmers in order to avoid unnecessary risks. The use of deep learning, data mining, satellites, drones, GPS maps of agricultural fields improve the performance of the resource use an increase the decision-making accuracy. Studies have shown that AI systems provide the farmers with knowledge and services which can increase their profitability. Farm sensors daily produce a huge volume of data which they cannot be processed by humans. AI agricultural systems generate a series of decisions for every occasion may occur and adapt them according to real time data [8].

AI technologies provide precision and accuracy in farmers operations. Farmers are able to improve their decision on what product they should produce, when this should happen and how much it would finally cost. With all the new smart systems the accurate decision-making can predict the amount of resources needed, the possible annual production, potential diseases, potential weather anomalies, the exact portions of fertilizers and pesticides from maximum production and minimum cost [8].

Farmers may use AI systems to increase productivity and agricultural precision and create forecasting patterns to avoid any production loss of disaster. The AI models are reproducing their patterns based on the incoming data. Thus, farmers are capable of knowing in real time the weather conditions, storms, hazel, and other extreme weather events. In undeveloped and developing countries where the incomes are modest, accurate weather and production forecasting could be proven a lifesaving technology. Sophisticated deep learning model could eventually provide a stability even to small farms where it is a fact that the daily generated data are small in volume. AI is considered as an integra part of the modern agriculture [60].

The precision agriculture is already involving a significant volume of data collection and processing aiming to generate AI models that will be able to increase the decision-making accuracy at the farm level. Referring to dynamically planting, spraying, and harvesting fields, it will prevent at an early-stage possible diseases and pests, and it will fertilize the crops when it is needed, and it will eventually increase the yields and profits. This new AI agents is expected to minimize the risk of food loss. By selectively providing different predictions for each farm, it promises to improve the resource performance of the farming industry, reducing water usage, controlling fertilizers, and pesticides, preventing them from uncontrollable misuse with serious drawbacks for humanity [73].

Based on the World Economic Forum research about harnessing artificial intelligence for the earth, it suggests that the 4th Industrial Revolution will include AI technologies combining robot engagement, chatbots communications for real time instructions, drone monitoring systems, synthetic biology, and new age materials.

Predictive analytics, deep learning, satellites, drones, internet of things (IoT), internet of behavior (IoB), mobile phones, and networks and will operate as one super AI entity. Agricultural data will be daily generated including information about aridity, moisture, temperature, and soil providing AI systems the ability to respond to these events instantly and dynamically and with the appropriate data to enhance production [73].

However, the excess use of personalized data without authorization generates other big privacy policy issues which need to be addressed because along with great technological advancement come great responsibilities. Artificial intelligence applications in agriculture will transform the way farmers, industries and public conceive it agriculture management. New technologies have the abilities to reshape bioeconomy in an excessive way, probably much more that the industrial revolution has ever done. The increase of the earth's population leads bioeconomy to a more controlling, optimizing and dynamically readjusting reality [73].

7.4.1 Examples of AI Applications in Agriculture

Smart Agriculture and IoF technologies use data and ML models to create AI patterns aiming to improve the quality of food through the production stages. As technology progresses and sensors become affordable, the volume of data collection increases providing real time information. Agriculture is benefited by the increase of data collection, storage, analysis, and modeling [57].

Agricultural robots are used to handle fundamental agricultural tasks such as spraying, fertilizing, monitoring, scanning, and harvesting crops. Robots are time efficient allowing farmers to be productive (Fig. 7.10) [45].

Sensor data analytics refers to factors like soil condition, weather forecast, seeds variety, diseases data which are constantly collected providing essential insights which will be used for future strategy. Based on the conditions, the price, the demand, farmers take better decisions (Fig. 7.10) [2].

Food safety lies on the base of continuous data collection goes the sensing techniques with hyper spectral imaging (HIS) and three dimensions laser scanning thousands of square meters of crops. Farmers are able to effortless monitor and intervene

Fig. 7.10 AI applications in agriculture (Reproduced from Javat Point [30])

into their farms through IoF technologies. IoF can monitor crops along their lifecycle from the planting until packaging [2].

Livestock management referring to crop and soil monitoring using image detection, image processing, satellite Imagery for geo-analytics imaging and machine learning algorithms to monitor crop status. Satellite images can show which regions need more irrigation and which have plenty. This leads to less waste in irrigation resources (Fig. 7.10) [45].

Crop readiness identification refers to ML image detection patterns which distinguish the green fruits into categories based on whether they are ripe enough to be sent for sale (Fig. 7.10) [2].

Predictive analytics are the very core particles of AI application on IoF. Without the data mining techniques and ML models, to track and predict the impact of environmental factors on crops, smart agriculture could not exist (Fig. 7.10) [45].

Disease detection refers also to ML image detection patterns which can scan, identify, and process plants images and determine whether they are diseased or they need fertilization or pesticides (Fig. 7.10) [2].

Water management refers to water conservation as it is becoming increasingly vital source of life as the world's population increases. AI algorithms provide farmers with a formula to control water consuming indicating possible water leaks, optimizing irrigation systems and scheduling. A ML system can optimally predict the specific amount of water needed (Fig. 7.10) [29].

Weather forecasting provides the farmers with the weather information of the same day and the following days with details such as humidity, temperature, etc. (Fig. 7.10) [62].

7.4.2   National Initiatives of AI in Agriculture

Artificial intelligence is considered a priority of policy agendas for governments around the world and stakeholders at both national and international levels. When it comes to bioeconomy certain countries have made a significant progress towards the AI national strategy and regulation regarding the investments, research, and implementation in agriculture when the venture capitals (VC) investors seem to have underestimate the role of AI in agriculture. Disregarding how important role sustainable systems play for the environment, VC investments in agriculture remain low in budget [46].

Among other industries including mobility and autonomous vehicles, healthcare, drugs and biotechnology, media, social platforms, marketing, IT infrastructure and hosting, business processes and support services, financial and insurance services, robots, sensors, information technology (IT) hardware, logistics, wholesale and retail, and digital security AI investments in agriculture ranks in the last position (Fig. 7.11) [46].

The global population continues to increase, generating a greater demand for food supply. It is estimated that between 1960 and 2016 the global population has

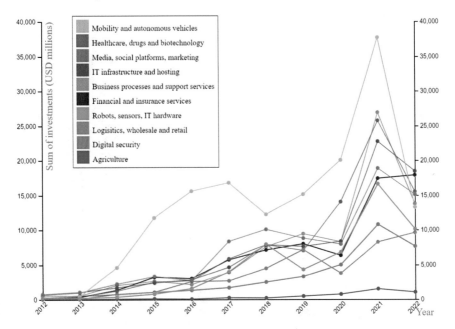

Fig. 7.11 VC investments in AI by industry (Reproduced from OECD.AI [46])

doubled. The need for bigger food production has emerged and only agriculture can efficiently address this issue [17].

Agriculture plays a significant role in increasing the national employment rates contributing to the national gross domestic product. According to the World Bank data, almost 37% (5 billion hectares) of total land surface currently used for crop production [72].

Despite that agriculture, has been at last position of AI investments it seems that it has set the pace for digital development. Along with the AI expansion, agriculture has started to play a significant role in restoring the environment and create the expectations of a more sustainable future [32].

Among the top countries with the biggest investments in AI for agriculture are the USA, China, EU, UK, Israel, India, Canada, Singapore, Japan, and Korea (Fig. 7.12) [47].

According to the International Telecommunication Union report the Food and Agriculture Organization (FAO) has initiated eight projects on AI including [31]:

Project 1: Strengthening global access to agricultural information and knowledge (Hand-In-Hand Geospatial Platform).
Project 2: Crop phenology and crop calendar with remote sensing and GEO-AI.
Project 3: Global and Country ASIS (Agriculture Stress Index System).
Project 4: FAO Data Lab.
Project 5: Detecting Fall armyworm through user submitted photos (FAMEWS).
Project 6: FAO Digital Portfolio.

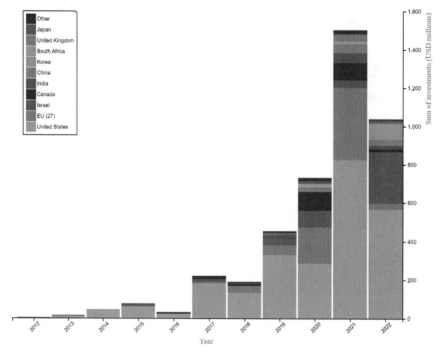

Fig. 7.12 Top countries in VC sum of investments in AI for agriculture (Reproduced from OECD.AI [47])

Project 7: iSharkFin (Identifying shark species from a picture of the fin).
Project 8: WaPOR (Water Productivity through Open access of Remotely sensed derived data).

7.4.2.1   AI Initiatives in Agriculture in United States of America (USA)

This part presents some of the best countries initiatives regarding the AI development. According to OECD.AI, USA as the leading country in artificial intelligence research, investments, and applications took an initiative referring to the application of AI systems in agriculture [48].

Directed by the National Science Foundation (NSF), and in partnership with the U.S. Department of Agriculture National Institute of Food and Agriculture (USDA), the U.S. Department of Homeland Security (DHS), Google, Amazon, Intel and Accenture, the National AI Research Institutes restores the connection in a broader nationwide network in order to pursue transformations in economic sectors, sciences, and engineering domains including food production and food safety. Among other objectives, is to support the AI-Driven Innovation in Agriculture and the Food System.

Led by the Federal Aviation Administration (FAA) a research program supports the safety of operations of Unmanned Aircraft Systems (UAS) enabling this way the development of UAS technologies for use among others, in agriculture promoting the development and enforcement of Federal regulations for UAS operations [48].

Let by National Institute of Food and Agriculture, an AI initiative called Food and Agriculture Cyber-informatics and Tools supported by big data, ML, autonomous systems, computer vision, bioinformatics, decision making systems, and socioeconomic. This initiative promotes and applies a on a wide variety of AI projects funding activities like agricultural systems and engineering, natural resources and environment, and Economics and rural communities. AI appliances refer to agricultural production process, sensors, bioinformatics, ecosystem sustainability, rural engagement, education, and training. Autonomous robots are expected to provide services in planting, yielding, fertilizing, monitoring, and harvesting. AI algorithms also use patterns for sustainable ecosystem management [42]

USA Departments of Agriculture service Agriculture Research Service (ARS) has developed a collaboration with industry to support AI application in agriculture. This collaboration includes livestock monitoring, harvesting robots, water conservation, pesticides management, using Unmanned Aerial Vehicle (UAV) technology. Economic Research Service (ERS) performs ML research and development for implementing better crop yielding models based on weather data. Through AI use, USDA brings high-quality food to the American family table [41].

The Agriculture and Food Research Initiative's Data Science for Food and Agricultural Systems (DSFAS) program focuses on connecting data science, AI, and agriculture. It aims to bring systems and communities close enough in order to know how to utilize data, understand resource management, and apply new AI systems to further US food and agriculture companies [43].

A new river forecast model called M^4 integrates AI for better water management. It will help farmers, ranchers, and foresters better organize their operations and provide for people worldwide who depend on American bioeconomy. The foundation of productive, sustainable agriculture also uses an AI forecast model for better water management [18, 19].

According to the USDA there are AI used cases which are already applied:

1. 4% Repair Dashboard
2. ARS Project Mapping
3. NAL Automated indexing
4. Forecasting Grasshopper Outbreaks in the Western United
5. States using Machine Learning Tools
6. Facial recognition
7. Coleridge Initiative
8. Show US the Data
9. Westat
10. Land Change Analysis Tool (LCAT)
11. Retailer Receipt Analysis
12. Ecosystem Management Decision Support System (EMDS)

13. Wildland Urban Interface - Mapping Wild-fire Loss
14. National Land Cover Database (NLCD) Tree Canopy Cover Mapping
15. The BIGMAP Project
16. DISTRIB-II: Habitat Suitability of Eastern United States Trees
17. CLT Knowledge Database
18. RMRS Raster Utility
19. TreeMap 2016
20. Landscape Change Monitoring System (LCMS)
21. Forest Health Detection Monitoring
22. Land Cover Data Development
23. Cropland Data Layer
24. List Frame Deadwood Identification
25. Climate Change Classification NLP
26. Video Surveillance System
27. Acquisition Approval Request Compliance Tool
28. Operational water supply forecasting for western US rivers [70].

Led by the OPEN (Open, Public, Electronic, and Necessary) Government Data Act, the Farm Service Agency (FSA) provides online stored data [21].

7.4.2.2   AI Initiatives in Agriculture in China

The Chinese government has taken AI initiative to promote smart agriculture and help the farmers, the producers and the tech companies co-exist and communicate. In order to success that it has increased the government funding for new digital essential technologies including AI, 5G, and IoF to ensure that the tech companies and farmers will have the least essential technological services available to start integrating AI systems. Services like cloud farming: high-tech assistant for precision agriculture, technical services for farmers, and remote sensing and drone assisted agricultural insurance have also been developed. The new national Chinese AI strategy implies the development of new research centers, open databases, and training. The government provides lower taxes, and loans to farmers [10].

Chinese e-government system for agricultural services provides limited online information and register. It provides adequate services regarding the advertisement and user pay system and restricted area and information sharing. It lacks privacy, security, disability, and foreign language services [34].

Regarding the public services, China has incorporated new technologies like big data, AI and 5G into e-government systems in order to promote the importance of electronic public sector services. In addition, the Chinese government plans to establish an online information resource system using blockchain technology. Social media are also smartly utilized to allow people, companies, and government to interact [69].

7.4.2.3   AI Initiatives in Agriculture in European Union (EU)

The European Commission promotes artificial intelligence research both in academia and industry including funding for investments in key projects society axes such as food and bioeconomy, health, education, industry, energy, internet technology. Automated driving, cyber-security, and e-governance. Through a series of funds supports the incorporation of AI/Robotic technologies in the European production line along with the education. The Commission also reinforces the creation of innovative market with pioneer services and products through a pilot program of the European Innovation Council which handles a budget of 2.7 billion Euros to support 1000 innovative projects and 3000 achieving awards. This pilot program can leverage AI systems development in many sections of public administration and industry [12].

Misuraca and van Noordt report mentions a research methodology which has been followed to identify which are the main challenges that the public sector faces from the use of AI. The entire process of AI application in the public sector allowed the researchers to record the upcoming AI strategies and practices across EU countries aiming to highlight the point of interest from AI incorporation in the public sector. There are 230 AI initiatives taken along with their characteristics, technological dimensions, specifications, and significance weights by the EU Member States. The explanatory analysis included a number of case studies were selected and thoroughly examined for finding AI initiatives in key research areas, such as agriculture, mobility and transport, social services, healthcare, and public administration. Th research also referred to AI and data governance case studies in a series of legal and administrative relevance. According to the report, AI can enable farmers to improve production rates; researchers can develop therapies for plant and animal diseases; it can control energy consumption by optimizing energy management systems; it can promote to a sustainable agriculture by decreasing the need for pesticides and water consumption; it can help provide analytical weather forecast and avoid catastrophic consequences for farmers [39].

Referring to the Geo-spatial and Earth/Observation project called Copernicus Program is the biggest earth observation and monitoring information system in the world. Regarding the agriculture economy and production, Copernicus program has embedded a free, full and open data policy which allows to monitor, collect, handle and provide huge volumes of structured data and processing performance through the advanced Data and Information Access Services (DIAS). The European Commission suggests, based on DIAS, to implement and apply AI systems and data from Copernicus to support and expand geo-location services for agriculture, soil quality, air quality, carbon footprint, aquaculture, and water conservation. It also takes AI initiatives to reinforce the usage of Earth observation data in public and private sector [13].

It is worth mentioning that among the EU states, there is one country, Estonia, with the most robust AI national strategy which takes many AI Initiatives are, Germany, and Belgium [48].

Geo-location-based services support precision agriculture, IoF, and data insights from interconnected software and devises. Data analytics aim to provide farmers with insights allowing them to analyse real-time data like weather forecast, GPS tracking and coordinates, temperature, moisture, growth, demand, prices and provide insights on how to optimize and increase yield. Data acquisition allows farmers to have an overview of their crops helping them to increase production quantities, modest risks, estimate the resources needed and implement an efficient production strategy [15].

7.4.2.4   AI Initiatives in Agriculture in United Kingdom (UK)

AI is essential to securing productivity, profitability, or sustainability in UK's agriculture economy in UK. The Department of Agriculture, Environment, and Rural Affairs (DAERA) controls a variety of environmental and productive pillars of the British economy [66]. The production and food sustainability and development, the ecosystems management, the aquaculture and fishing industry, and the forestry conservation in regard to the ethical policy, human values, animal protection, and the environment conservation. DAERA engaged AI companies to assess and provide solutions to a series of issues regarding the precision agriculture, planting, yielding, and animal tracking [44].

7.4.2.5   AI Initiatives in Agriculture in Israel

Israel Innovation Authority (IIA) refers to an ongoing Planned AI R&D Framework & Activities which includes activities like the Joint National Infrastructure for Computing and Storage, free food and agriculture datasets access, research laboratories and investors to support research in plant inspection, it also supports training and development and it supports new entrepreneurs consotiums for the future challenges [48]. The state of Israel is one of the most advanced states in AI applications in agri-food business in the world. It applies AI systems in 3 axes of agri-tech development including AI in agriculture, robotics and sensors. Speaking of AI in agriculture, satellite images, aerial photos, big data collection, and ML algorithms provide fully operational decision making system which monitors the fields and provides farmers with insights. When it comes to robotics and sensors equiped with computer vision they are able to recognize, monitor and pick fruits from the trees [27].

7.4.2.6   AI Initiatives in Agriculture in India

AI in agriculture in India is expected to increase at a rate of 22.5% reaching 2.6 billion USD by the year of 2025. AI currently provides farmers with services which help them to acquire higher rate of yield quantities through optimized recommendation about the quality of crop, the seeds, and the natural resources. Machine learning generates predictive models which help farmers organize their annual production strategy [6]. The government of India has started to leverage AI to help deliver government services to the citizens. The main application of AI in agri-food business sector in India has been in the domain of machine learning and data mining. The international organization of International Crops Research Institute for the Semi-Arid Tropics (ICRISAT) has developed a collaboration with Microsoft to develop an AI Sowing application using ML and Power BI. It provides suggestions to farmers at their mobiles about the optimal date to sow. The government has established a memorandum of understanding with Microsoft to use predictive analytics for predicting the commodity pricing [4].

7.4.2.7 AI Initiatives in Agriculture in Canada

AI application in agriculture has transformed Canada's agri-food sector enhancing Canada's position worldwide. The Canadian government has invested 3 billion USD to a five-year AI investment to help Canada's bioeconomy grow, overcome unstable weather challenges, and increase profits. In order Canada to be among the strongest and fully agricultural developed countries the department of Agriculture and Agri-Food, has collaborated with companies under the AgriScience Program. The companies have understood the major drawbacks of farming sector and they will develop an AI data recording kit based on sensor technology and predictive analytics. This application will collect GPS and camera data in order to create representation of the farms in order to be able to provide suggestion to farmers when it is required. AI agriculture has altered the way farmers work. AI agri-food systems automatically monitor, collect, process, and suggest providing accuracy, efficiency, sustainability, and increased productivity [1].

7.4.2.8 AI Initiatives in Agriculture in Singapore

According to the Singapore Food Agency, the government has invested 30 million USD to a project called "30 × 30 Express" to promote the agriculture industry to upgrade the local eggs, leafy vegetables, and fish production as quickly as possible. The Singapore government has established a promising plan called '30 by 30', to manage produce 30% of its country's food needs by 2030. It enables AI robots for harvesting kale and nurturing the local food of barramundi. The government of Singapore aims to increase the level of farming in the country and utterly promote the countries products [61].

Hence, is has engaged several companies to work with the government on a series of AI project. Companies conduct research and apply innovative AI techniques in farming and fisheries. Some of their services include smart sensors use to detect essential to the plants, fish, and food parameters such as temperature, humidity, and light levels. All the parameters are automatically readjusted upon the environment indications and the product properties. There is also a local farm of different kinds of fish which uses its producing inhouse its own software for water monitoring and fish condition aiming to maintain the current population healthy and gradually increase it. A local farm manages to mimic Japan's climate in order to grow Japanese crops like Karashina leaves and Wasabina mustard greens. This environmental adaptation is based on AI sensors and advanced decision-making systems which control the contained farm aiming to apply the same or similar environment characteristics. Since Singapore lacks vast farmlands, local companies have surpassed the obstacles by allowing their crops to grow on the roofs of building. Hydroponic farming has also been affected by the AI agriculture. A local farm produces kale, arugula, and strawberries in a fully controlled artificial environment using artificial lighting [61].

Regarding the public good, the government of Singapore has incorporated smart technologies to support of the smart national strategy including the government portal (Gov.sg) that provides access to e-participation (reach.gov.sg), e-services (citizen-connectcentre.sg), open data (data.gov.sg), and public procurement (gebiz.gov.sg). The government has also invested in effort and funds in cybersecurity [69].

7.4.2.9   AI Initiatives in Agriculture in Japan

The Japanese government has designed a national AI strategy that promotes the use of AI to the public sector. Part of this AI strategy refers to bioeconomy. AI in agriculture is part of a smart agriculture initiative. The Government of Japan considers AI a core part of its growth strategy in seeking to achieve a new form of capitalism [67].

Since 2019, a national strategy has been in motion to conduct R&D and implement AI in society. Under the AI Strategy 2022 announced this April, Japan aims to create Society 5.0—the next logical step in the progression from a hunter-gatherer society through an agricultural society, industrial society, and the information society of today—which will achieve advanced convergence between cyberspace and physical space. In order to take that step, the AI strategy sets new goals to make AI a more entrenched part of society and creates a concrete action plan to respond to emerging threats and challenges facing society, such as pandemics and major disasters. Undoubtedly, the AI has reshaped the scenery of agriculture [67].

However, there are obstacles which must be overcome. Japan's climate is hospitable for many pests which are designated as harmful species by the government. Earth changing environment creates new climate conditions work in favor of the pests' populations. In order to help farmers, efficiently identify a variety of pests, the National Agriculture and Food Research Organization (NARO) has implemented an AI system using predictive analytics. NARO decided to use a cloud service, to release a web Application Programming Interface (API) to provide companies the ability to build their pest applications opening the practical adoption of an AI system in real agricultural situations [67].

In Japan, the Government Digital Transformation Plan focuses on new technologies to enhance electronic governance and improve the citizens online interactions. The Council for Science, Technology and Innovation takes e-government initiatives aiming to support business development in Japan. Digital services include platforms like the central portal for digital government (e-gov.go.jp), e-participation platforms (e-Testimony), open data (data.go.jp), public procurement platforms (geps. go.jp), and the e-government legal platform [69].

7.4.2.10   AI Initiatives in Agriculture in South Korea

The Republic of South Korea is recognized as one of the best world's technologically advanced countries in information and communication technology including areas such as telecommunication, digital transformation, and e-Government services [11].

Based on the OECD.AI report, the Korea government has taken fourteen AI initiatives in an effort to digitally transform the public sector services [49].

According to Lee et al. [34], the Korean government is among the countries with the most well organized and robust government agricultural websites. It should be mentioned that there is the need for greater focus should be given on advanced functions including disability, data privacy, users' security, advertisements, and user payment systems.

In the space of 60 plus years, Korea has transformed from an agricultural country to one high-tech country with automobiles, shipbuilding, advanced manufacturing, smart constructing, and advanced digital agricultural with molding systems that assist farmers with their production [40].

The transformation of agriculture into digital agriculture engaged several South Korean public institutions which have prototyped in the data-driven agriculture ecosystem. Institution like Korea Agency of Education, Promotion and Information Service in Food, Agriculture, Forestry and Fisheries (EPIS) which is a state agency under the Ministry of Agriculture, Food and Rural Affairs (MAFRA), which leads the information and communications technology (ICT) knowledge promotion in rural areas enabling human resources for agriculture and aquaculture. EPIS has contributed to the development for "Smart Farm Innovation Valley". The latter operates as an investor and agriculture data repository to help young agriculture entrepreneurs grow their businesses and enhance AI applications in agriculture [51].

Since the agricultural population in Korea is ageing along with the climate change and environmental issues an immediate response is required. Smart farming, entrepreneurship, ICT innovation, and fund raising are required on the "Smart Farm Innovation Valley in Gimje, Jeollabuk-do", enabling synergies among farmers, agri-businesses, and research centers [64].

Smart agriculture will increase food safety and productivity through data-based decision making. Incubators, industrial ecosystems, big data, livestock and upland farming are used for agri-business growth [22].

7.5 Discussion

Regarding AI it is panacea to think that "logical choice" is to choose the action with the highest expected performance because the human choices consist of complex decision-making factors and processes in real-world [68].

There are seven recommendations which will help governments use of AI decision-making systems safely, sustainably, and ethically: Test before use to avoid any unexpected outcomes or problems. Fair services for all users and citizens. Define responsibilities. Protect the safety of citizens data. Educate users and citizens about the advantages and disadvantages of AI. Ensure that AI projects follow the law. Design an expandable system that will avoid potential future consequences. The ethical issues of artificial intelligence applications are at the center of technological progress in human society. Recent studies mentioned that people have no trust for advanced technologies regulations [26].

According to OECD.AI, the principles represent the proper use of innovative and trustworthy AI technology with respect to the human rights. OECD adopts a series of AI principles that can be practical and flexible enough to endure the challenges of the new AI era. The AI principles are distinguished in Values-based principles and recommendations for policy makers [47].

The values-based principles are the following: Inclusive growth, sustainable development and well-being, inclusive growth, sustainable development and well-being, transparency and explainability, robustness, security, and safety, and ac-countability. The recommendations for policy makers are the following: Investing in AI R&D, fostering a digital ecosystem for AI, providing an enabling policy environment for AI, building human capacity, and preparing for labor market transition, international co-operation for trustworthy AI.

The use of big data and AI systems in the bioeconomy offer significant benefits to agriculture and can address the upcoming production and climate challenges, the increasing food demand, and privacy concerns. AI must reassure the citizens that the information will be ethically and responsibly handled, and government will commit to AI transparency and accountability through good governance practices and laws.

The challenge is to scale up the world's commitment to AI, so the governments worldwide to unlock the opportunity AI has to offer, not just in the bioeconomical and agricultural sectors but also in public sector, hospitals, houses, universities, and industries.

Governments around the world, especially the wealthiest, play an important role in creating the conditions needed in which safe and fair AI will be used for bioeconomy aiming to foresee the amount of production needed in the future in order to eliminate the famine from poor counties, increase the food safety and quality, control water waste, perform fertilizing and pesticides efficient management, and minimize the environmental footprint.

7.6   Conclusion

In a world of big data, autonomous robots, smart mobiles, and cameras the expansion of computational power in inevitable. The landscape of information management and privacy has drastically altered. People's lives are interconnected through IoT devices, mobiles, and smart cities applications. AI potential promises wealth of potential benefits, including more dynamic use planet resources, enhanced decision-making, and lower carbon emissions. Fueled by AI technology the bioeconomy including agriculture, aquaculture, and forestry, the health system, the judicial system, the education, and the e-government services can be reenforced. Governments around the world try to respond to the emerging climate change and they adapt AI technologies to their national state strategy. States with strong agricultural sector apply AI systems to optimize the profits and resources management and minimize the expenses and energy. Based on the latest technological advances in AI, government agencies have initiated programs that use new technologies to optimize their system performance and citizens satisfaction. However, the lack of AI experts, computational efficient systems, AI trustworthiness, interoperability and AI interpretability consists actual barriers and challenges towards the AI adoption.

Since the beginning of twenty-first century e-government systems arrival decreased the time needed for information retrieval by directly providing the information required to the users through the e-government applications. Application such as online services, websites, foreign language access, personalized data and support centers, data visualization, weather forecast, farming recommendations, available funding projects, access for people with disabilities, online payment systems, privacy and information sharing have led governments to continue upgrading the e-government services.

Referring to the users, farmers are also included. The days when farmers relied on what government told have irreversibly passed. Farmers are currently adapted to a new technology reality of information dissemination engaging smart services, devices, e-learning and social media communication. Digital communication allows government increase farmers satisfaction and collect data at the same time for pattern generation, decision support systems optimization and collaboration. Through a top-down communication, farmers are currently able to interact and share information with each other and with the government systems in real time using their mobiles. Farmers are not only connected to the digital world, but they also help to their life's development. Despite that the least technologically advanced countries face severe infrastructure problems it is essential to introduce the information and communication technology in order to build the foundations of a technologically sustainable future for qualitative e-government services.

During the information evolution era, the e-government initiatives taken by countries around the globe have been successful. Governments have imported a series of measured taken for public sector reform. E-government initiatives based on AI technology manage to provide more user centric and personalized services based on sophisticated systems and data storage, data management and data mining. A new e-governance perspective based on AI established a new position in the fourth industrial revolution, which refers to decentralized self-service model in a fully autonomous network and a personalized platform.

However, there are still several barriers to the wide implementation scheme of AI systems in bioeconomy and agriculture. Due to the initial AI investment costs and lack of training referring to the AI systems performance and despite that AI provide many benefits and optimizations, many farmers and producers seem to have adopted an attitude of patience until they feel confident to proceed. Thus, bringing together the different stakeholders in global agriculture to take the challenge of promoting the benefits of technology benefits implies that some of the obstacles which farmers regularly face will be quickly overcome without leaving farmers exposed to certain business risks. Most of the financially biggest countries have understood the importance of using AI systems in order to optimize procedures towards a sustainable agri-food business sector and a stable environment low in carbon emissions. Yet, AI generates social, technological, human rights, legal challenges, and ethical considerations.

In an effort to attribute the role of artificial intelligence application in the e-government systems in bioeconomy and especially in agriculture the authors have tried to research and collect data which indicate the actual stage of AI participation into the public sector. Data was collected from the wealthiest and the most technologically advanced countries around the world. The current research methodology aimed to provide an introductory analysis of the AI terms and growth across the industrial sectors, the AI current value, the AI system model, the countries AI readiness, the e-government system or portal using AI technology, the big data contribution, the IoF government model for bioeconomy, the investments in AI across sectors, the AI applications in agriculture including the major AI government initiatives for agriculture around the globe. The scope of this chapter is to gradually introduce the terms and notions to the people who actually need to make a step forward in AI appliances in bioeconomy.

Referring to the future research, an assessment framework is introduced which will be used in a future research attempt to measure the AI readiness level of public agri-sector among the most technologically advanced counties. This framework will manage to statistically record and export results about the countries readiness in AI technology in agriculture enabling researchers draw conclusions. This framework will measure the existing e-government services which use AI among the wealthiest countries aiming to tackle the significance of AI in the overall economy, the agricultural growth, and the farmers satisfaction. Through a series of statistical formulas, it will provide an overview of the countries AI progress, potential AI benefits, and drawbacks. The countries public agri-sector readiness will be classified according to the level of service provision to the public. This research will be a referral point for the governments which will decide to mimic the countries which have moved ahead in AI adoption setting the foundations for a sustainable economy and environment.

References

1. Agriculture and Agri-Food Canada. (2022). Government of Canada invests in digitization of farming to strengthen sustainability of Canada's agriculture sector. Retrieved July 28, 2022, from https://www.canada.ca/en/agriculture-agri-food/news/2022/04/government-of-canada-invests-in-digitization-of-farming-to-strengthen-sustainability-of-canadas-agriculture-sector.html
2. Bagchi, A. (2018). Artificial intelligence in agriculture. Retrieved July 15, 2022, from https://www.mindtree.com/sites/default/files/2018-04/Artificial%20Intelligence%20in%20Agriculture.pdf
3. Bansal, M., Sirpal, V., & Choudhary, M. K. (2022). Advancing e-government using internet of things. In S. Shakya, R. Bestak, R. Palanisamy, K. A. Kamel (Eds.), *Mobile Computing and Sustainable Informatics. Lecture notes on Data Engineering and Communications Technologies* (p. 68). Springer. https://doi.org/10.1007/978-981-16-1866-6_8
4. Basu, A., & Hickok, E. (2018). Artificial intelligence in the governance sector in India. Retrieved August 4, 2022, from https://cis-india.org/internet-governance/ai-and-governance-case-study-pdf
5. Berryhill, J., Kok Heang, K., Clogher, R., & McBride, K. (2019). Hello, world: Artificial intelligence and its use in the public sector. OECD Working Papers on Public Governance, No. 36, OECD Publishing, Paris. https://doi.org/10.1787/726fd39d-en
6. Bordoloi, P. (2022). *India's Digital Agriculture Mission is About People, Not Projects.* Retrieved August 4, 2022, from https://analyticsindiamag.com/indias-digital-agriculture-mission-is-about-people-not-projects/
7. Charvat, K., Charvat, K. Jr., Reznik, T., Lukas, V., Jedlicka, K., Palma, R., & Berzins, R. (2018). Advanced visualisation of big data for agriculture as part of databio development. In *IGARSS 2018—2018 IEEE International Geoscience and Remote Sensing Symposium* (pp 415–418). https://doi.org/10.1109/IGARSS.2018.8517556
8. Charvat, K., Horakova, S., Rogotis, S., Catucci, A., Auran, P., Poulakidas, A., & Habyarimana, E. (2017). DataBio—D1.1—Agriculture pilot definition. *European Journal of Agronomy, 29*(2–3), 59–71. https://doi.org/10.13140/RG.2.2.12113.63847
9. Chui, M., Manyika, J., Miremadi, M., Henke, N., Chung, R., Nel, P., & Malhotra, S. (2022). *Notes from the AI Frontier: Applications and Value of Deep Learning.* Retrieved July 10, 2022, from https://www.mckinsey.com/featured-insights/artificial-intelligence/notes-from-the-ai-frontier-applications-and-value-of-deep-learning
10. Chun, S. (2019). *How AI is Modernizing Chinese Agriculture.* Retrieved July 20, 2022, from https://medium.com/syncedreview/how-ai-is-modernizing-chinese-agriculture-9248423c0b8a
11. Chung, C.-S., Choi, H., & Cho, Y. (2022). Analysis of digital governance transition in South Korea: Focusing on the leadership of the president for government Innovation. *Journal of Open Innovation Technology, Market, and Complexity, 8*(2). https://doi.org/10.3390/joitmc8010002
12. European Commission. (2018). *Artificial Intelligence for Europe.* Retrieved July 28, 2022, from https://ec.europa.eu/jrc/communities/sites/default/files/communicationartificialintelligence.pdf
13. European Commission. (2018). *Coordinated Plan on Artificial Intelligence.* Retrieved July 28, 2022, from https://ec.europa.eu/jrc/communities/sites/default/files/annexenglish.pdf
14. European Commission. (2020). *White Paper on Artificial Intelligence—A European Approach to Excellence and Trust.* Retrieved July 22, 2022, from https://ec.europa.eu/info/sites/info/files/commission-white-paper-artificial-intelligence-feb2020_en.pdf
15. European Commission. (2022). *Data Act.* Retrieved August 2, 2022, from https://digital-strategy.ec.europa.eu/en/policies/data-act
16. European Commission (2018). Directorate-General for Research and Innovation: A sustainable bioeconomy for Europe: Strengthening the connection between economy, society and the environment: updated bioeconomy strategy, Publications Office. Retrieved July 24, 2022, from. https://doi.org/10.2777/792130

17. FAO. (2020). *Land use in Agriculture by the Numbers*. Retrieved July 28, 2022, from https://www.fao.org/sustainability/news/detail/en/c/1274219/
18. Farmers. (2022). *New River Forecast Model Integrates Artificial Intelligence for Better Water Management in the West*. Retrieved August 20, 2022, from https://www.farmers.gov/blog/new-river-forecast-model-integrates-artificial-intelligence-better-water-management-in-west
19. Farmers. (2022). *Soil Health*. Retrieved August 20, 2022, from https://www.farmers.gov/conservation/soil-health
20. Feldstein, S. (2019). *We Need to Get Smart About How Governments Use AI*. Retrieved July 08, 2022, from https://carnegieendowment.org/2019/01/22/we-need-to-get-smart-about-how-governments-use-ai-pub-78179
21. FSA. (2022). *FSA Online Data Resources*. Retrieved August 27, 2022, from https://www.fsa.usda.gov/online-services/index
22. G20 Information Center. (2019). *Establishment of Smart Farm Innovation Valley*. Retrieved August 7, 2022, from http://www.g20.utoronto.ca/2019/2019-g20_niigata_yuuryou-8.pdf
23. Ganeshkumar, C., Jena, S. K., Sivakumar, A., & Nambirajan, T. (2021). Artificial intelligence in agricultural value chain: Review and future directions. *Journal of Agribusiness in Developing and Emerging Economies*. https://doi.org/10.1108/jadee-07-2020-0140
24. GDS. (2020). *A Guide to Using Artificial Intelligence in the Public Sector*. Retrieved July 28, 2022, from https://assets.publishing.service.gov.uk/government/uploads/system/uploads/attachment_data/file/964787/A_guide_to_using_AI_in_the_public_sector__Mobile_version_.pdf
25. Google Trends. (2022). *Artificial Intelligence*. Retrieved August 20, 2022, from https://trends.google.com/trends/explore?date=2012-07-16%202022-08-16&q=artificial%20intelligence
26. GOV.UK. (2021). *Guidance Ethics, Transparency and Accountability Framework for Automated Decision-Making*. Retrieved July 28, 2022, from https://www.gov.uk/government/publications/ethics-transparency-and-accountability-framework-for-automated-decision-making/ethics-transparency-and-accountability-framework-for-automated-decision-making
27. Guide2Agricutlure. (2020). *Top 7 Agricultural Technologies used in Israel*. Retrieved July 27, 2022, from https://guide2agriculture.com/author/prakash/
28. Heeks, R. (2005). *Foundations of ICTs in Development: The Onion-Ring Model*. https://doi.org/10.13140/RG.2.2.24441.01127
29. Hunt, S. (2021). *How the Agriculture Industry is Using AI*. Retrieved July 27, 2022, from https://www.datamation.com/artificial-intelligence/ai-in-agriculture
30. Javat Point. (2021). *Artificial Intelligence in Agriculture*. Retrieved July 19, 2022, from https://www.javatpoint.com/artificial-intelligence-in-agriculture
31. International Telecommunication Union. (2021). *United Nations Activities on Artificial Intelligence (AI)*. Retrieved August 27, 2022, from https://www.itu.int/dms_pub/itu-s/opb/gen/S-GEN-UNACT-2021-PDF-E.pdf
32. Kundalia, K., Patel, Y., & Shah, M. (2020). Multi-label movie genre detection from a movie poster using knowledge transfer learning. *Augmented Human Research*, *5*(11). https://doi.org/10.1007/s41133-019-0029-y
33. Lane, J. (2021). *Artificial Intelligence & Machine Learning in Bioeconomy: The Digest's 2020 Multi-slide Guide to Idaho National Laboratory*. Retrieved July 24, 2022, from https://www.biofuelsdigest.com/bdigest/2021/01/14/artificial-intelligence-machine-learning-in-bioeconomy-the-digests-2020-multi-slide-guide-to-idaho-national-laboratory/4/
34. Lee, T.-R., Wu, H.-C., Lin, C.-J., & Wang, H.-T. (2008). Agricultural e-government in China, Korea, Taiwan and the USA. *Electronic Government, an International Journal*, *5*(1), 63. https://doi.org/10.1504/eg.2008.016128
74. Leslie, D., Burr, C., Aitken, M., Cowls, J., Katell, M., & Briggs, M. (2021). *Artificial intelligence, human rights, democracy, and the rule of law: a primer*. The Council of Europe. Retrieved July 15, 2022, from https://rm.coe.int/primer-en-new-cover-pages-coe-english-compressed-2754-7186-0228-v-1/1680a2fd4a
35. Manzoni, M., Medaglia, R., Tangi, L., Van Noordt, C., Vaccari, L., & Gattwinkel, D. (2022). *AI Watch. Road to the Adoption of Artificial Intelligence by the Public Sector*. Publications office

of the European Union, Luxembourg, ISBN 978-92-76-52132-7, JRC129100. https://doi.org/10.2760/288757

36. Medaglia, R., Gil-Garcia, J. R., & Pardo, T. A. (2021). Artificial intelligence in government: Taking stock and moving forward. *Social Science Computer Review*. https://doi.org/10.1177/08944393211034087

37. Mehr, H. (2017). *Artificial intelligence for citizen services and government. Harvard Ash Center Technology & Democracy Fellow*. Retrieved June 17, 2022, from https://ash.harvard.edu/files/ash/files/artificial_intelligence_for_citizen_services.pdf

38. Misuraca, G., Van Noordt, C., & Boukli, A. (2020). The use of AI in public services: results from a preliminary mapping across the EU. In *Proceedings of the 13th International Conference on Theory and Practice of Electronic Governance (ICEGOV 2020)*. Association for Computing Machinery, New York, NY, USA (pp. 90–99). https://doi.org/10.1145/3428502.3428513

39. Misuraca, G., & van Noordt, C. (2020). Overview of the use and impact of AI in public services in the EU, EUR 30255 EN, Publications Office of the European Union, Luxembourg. ISBN 978-92-76-19540-5. https://doi.org/10.2760/039619, JRC120399

40. Myeong, S. (2019). E-government to smart e-governance: Korean experience and challenges. In: A. Farazmand (Ed.), *Global Encyclopedia of Public Administration, Public Policy, and Governance*. Springer. https://doi.org/10.1007/978-3-319-31816-5_3814-1

41. NAII Strategic pillars. (2022). *Applications*. Retrieved July 22, 2022, from https://www.ai.gov/strategic-pillars/applications/

42. NIFA. (2022). *Artificial Intelligence*. Retrieved July 25, 2022, from https://www.nifa.usda.gov/artificial-intelligence

43. NIFA. (2022). *Data Science for Food and Agricultural Systems (DSFAS)*. Retrieved July 28, 2022, from https://www.nifa.usda.gov/grants/programs/data-science-food-agricultural-systems-dsfas

44. Objectivity. (2022). *Using AI to Modernise Agricultural Processes*. Retrieved July 23, 2022, from https://www.objectivity.co.uk/case-studies/daera-using-ai-to-modernise-agricultural-processes/

45. OECD. (2019). Artificial Intelligence in Society. OECD Publishing. https://doi.org/10.1787/eedfee77-en

46. OECD.AI. (2022). *OECD VC investments in AI by industry. Visualisations powered by JSI using data from Preqin*. Retrieved July 28, 2022, from www.oecd.ai

47. OECD.AI. (2022). *OECD AI Principles overview. Visualisations powered by JSI using data from Preqin*. Retrieved July 28, 2022, from www.oecd.ai

48. OECD.AI. (2022). *National AI Policies & Strategies*. Retrieved July 25, 2022, from https://oecd.ai/en/dashboards

49. OECD.AI. (2021). *Powered by EC/OECD (2021), Database of national AI policies (2021)*. Retrieved August 20, 2022, from www.oecd.ai

50. OMNIA. (2022). *How is Automated IT Revolutionizing Government Agencies?* Retrieved July 03, 2022, from https://www.americancityandcounty.com/2022/02/24/how-is-automated-it-revolutionizing-government-agencies

51. Open Learning Campus Digital ag series. (2021). *Fostering Digital Agriculture Ecosystems and Smart Farming in Korea—Case of Smart Farm Innovation Valleys*. Retrieved August 28, 2022, from https://olc.worldbank.org/content/digital-ag-series-fostering-digital-agriculture-ecosystems-and-smart-farming-korea-case

52. OVIC. (2022). *Artificial Intelligence and Privacy—Issues and Challenges*. Retrieved July 09, 2022, from https://ovic.vic.gov.au/privacy/artificial-intelligence-and-privacy-issues-and-challenges

53. Oxford Insights. (2021). *The 2021 Government AI Readiness Index, 2021*. Retrieved July 19, 2022, from https://static1.squarespace.com/static/58b2e92c1e5b6c828058484e/t/61ead0752e7529590e98d35f/1642778757117/Government_AI_Readiness_21.pdf

54. Panpatte, S., & Ganeshkumar, C. (2021). Artificial intelligence in agriculture sector: Case study of blue river technology. In D. Goyal, A. K. Gupta, V. Piuri, M. Ganzha, & M. Paprzycki (Eds.), *Proceedings of the Second International Conference on Information Management and*

Machine Intelligence. Lecture Notes in Networks and Systems, Vol. 166. Springer. https://doi. org/10.1007/978-981-15-9689-6_17

55. Raji, B. S. (2022). Exploring how artificial intelligence (AI) can support start-ups to manage crisis situations for future sustainable business in the agri-food industry. In M. Ali (Eds.), *Future Role of Sustainable Innovative Technologies in Crisis Management* (pp. 192–213). IGI Global. https://doi.org/10.4018/978-1-7998-9815-3.ch014

56. Richards, M. (2022). *The Future of AI: Experts Weigh in on Global Competition and Fairness.* Retrieved July 10, 2022, from https://www.uschamber.com/technology/the-future-of-ai-experts-weigh-in-on-global-competition-and-fairness

57. Rodriguez, N. (2022). *AI Models that Drive Digital Transformation.* Retrieved July 16, 2022, from https://www.aimodellist.com

58. Samoili, S., Lopez Cobo, M., Gomez Gutierrez, E., De Prato, G., Martinez-Plumed, F., & Delipetrev, B. (2020). AI watch. Defining artificial intelligence, EUR 30117 EN, Publications office of the European Union, Luxembourg, ISBN 978-92-76-17045-7, JRC118163. https:// doi.org/10.2760/382730

59. Saura, J. R., & Debasa, F. (2022). *Handbook of Research on Artificial Intelligence in Government Practices and Processes.* IGI Global. https://doi.org/10.4018/978-1-7998-9609-8

60. Schellberg, J., Hill, M. J., Gerhards, R., Rothmund, M., & Braun, M. H. (2008). Precision agriculture on grassland: Applications, perspectives and constraints. *European Journal of Agronomy, 29,* 59–71. https://doi.org/10.1016/j.eja.2008.05.005

61. SFA. (2022). *Fertile futures - Agritech in SG.* Retrieved July 20, 2022, from https://www.sfa. gov.sg/fromSGtoSG/our-sg-food-story/fertile-futures-agritech-in-sg

62. Shandilya, U., & Khanduja, V. (2020). Intelligent farming system with weather forecast support and crop prediction. In *5th International Conference on Computing, Communication and Security (ICCCS)* (pp. 1–6). https://doi.org/10.1109/ICCCS49678.2020.9277437

63. Shimonti, P. (2018). *Primary Sector of the Economy.* Retrieved July 19, 2022, from https:// www.geospatialworld.net/entity/primary-sector-of-the-economy/

64. Smart City Korea. (2022). *Jeollabuk-do, Gimje smart farm innovation valley prepared a Korean-style smart farm operation model that converges ICT.* Retrieved August 13, 2022, from https://smartcity.go.kr/en/

65. Södergård, C. (2021). Summary of potential and exploitation of big data and AI in bioeconomy. In C. Södergård, T. Mildorf, E. Habyarimana, A. J. Berre, J. A. Fernandes, C. Zinke-Wehlmann (Eds.), *Big Data in Bioeconomy.* Springer. https://doi.org/10.1007/978-3-030-71069-9_32

66. Suthern, M., & Martin, R. (2021). *Insight to AI in UK agri-tech.* Retrieved July 24, 2022, from https://www.barclays.co.uk/content/dam/documents/business/business-insight/Insights_ AI_in_Agriculture.pdf

67. The Government of Japan. (2022). *The Impact of AI: Anyone can be a Skilled Farmer.* Retrieved July 27, 2022, from https://www.japan.go.jp/kizuna/2022/05/anyone_can_be_a_s killed_farmer.html

68. UNESCO. (2021). *Recommendation on the Ethics of Artificial Intelligence.* Retrieved July 19, 2022, from https://en.unesco.org/artificial-intelligence/ethics

69. United Nations. (2020). *E-government survey 2020. Digital government in the decade of action for sustainable development.* Retrieved August 20, 2022, from https://publicadministration.un. org/en/Research/UN-e-Government-Surveys

70. USDA. (2022). *Inventory of USDA Artificial Intelligence Use Cases.* Retrieved August 21, 2022, from https://www.usda.gov/data/AI_Inventory

71. World Bank. (2020). *Artificial intelligence in the public sector maximizing opportunities, managing risks.* Retrieved July 20, 2022, from https://openknowledge.worldbank.org/bitstr eam/handle/10986/35317/Artificial-Intelligence-in-the-Public-Sector-Maximizing-Opportuni ties-Managing-Risks.pdf?sequence=1&isAllowed=y

72. World Bank. (2022). *Agricultural land (% of land area).* Retrieved July 20, 2022, from https:// data.worldbank.org/indicator/AG.LND.AGRI.ZS?end=2018&start=1961

73. World Economic Forum. (2018). *Harnessing artificial intelligence for the earth report 2018.* Retrieved July 26, 2022, from https://www3.weforum.org/docs/Harnessing_Artificial_Intell igence_for_the_Earth_report_2018.pdf

Chapter 8
Enhancing Government Actions Against Covid-19 Using Computer Science

Christophe Gaie and Markus Mueck

Abstract After more than two years of fighting the Covid-19 pandemic, governments face the challenge of maintaining a high level of protection while a worldwide economic crisis appears and citizens are questioning and partially rejecting protection measures. As a consequence, it is essential to implement concrete measures in support of a Vaccination and Protection Strategy (VPS). This chapter defines a strategy based on complementary dimensions: (i) The Neutralization Antibody Sustaining Method; (ii) The Vaccination Fraud Detection Method; (iii) The School Blockchain Detection Method, and (iv) The Sanitary Measures Optimization. The fourth dimension is deeply explored as it details how to implement different innovations and proposes a comparison to define which proposal should be enforced urgently. The various measures are evaluated based on a common criteria approach.

Keywords e-Government · Data science · Data analytics · Artificial intelligence · Covid-19 · Pandemic · Vaccination

8.1 Introduction

The last few years have seen the emergence of multiple techniques to fight the Covid-19 pandemic through various approaches in terms of sanitary measures (protection measures, sanitary restrictions, medicine distribution and vaccine distribution) and in terms of scientific techniques (neural networks, resource allocation, genetic algorithms). A paper by Jordan et al. [1] proposes a comprehensive framework to classify the methods that aim to predict and control the pandemic.

In the literature, there are many methods for supporting the fight against the pandemic, such as facility management [2], vaccination roll-out [3], methods for

C. Gaie (✉)
Centre interministériel de services informatiques relatifs aux ressources humaines (CISIRH), 41 Boulevard Vincent-Auriol, 75013 Paris 13, France
e-mail: christophe.gaie@gmail.com

M. Mueck
Jaegerstrasse 4b, 82008 Unterhaching, Germany

© The Author(s), under exclusive license to Springer Nature Switzerland AG 2023   173
C. Gaie and M. Mehta (eds.), *Recent Advances in Data and Algorithms for e-Government*,
Artificial Intelligence-Enhanced Software and Systems Engineering 5,
https://doi.org/10.1007/978-3-031-22408-9_8

reducing the virus spread [4], lockdown approaches [5] or vaccine allocation targeting different age groups [6]. Although these methods are promising, they tackle the single dimension of protection while putting aside many other concerns such as transparency of the government actions or long-term acceptance of the impacts on the economy.

This chapter aims to offer new technologies and tools so that governments can not only maintain an adequate level of protection but also decrease the alertness of citizens. Indeed, there is very few research that addresses the vaccination and protection strategies in every dimension. Researchers only focus on a single aim at any one time: to maximize the vaccination coverage, improve citizen trust, or preserve the economy. They often model the problem in a single dimension without describing the concrete situation in concrete situations (hospitals, schools or supermarkets). Thus, the current research towards fighting Covid-19 and other pandemics is not immediately applicable and can only provide limited effects.

On the contrary, the authors believe that decision-makers have to define a strategy that can simultaneously address the problem concretely and entirely and be accepted by a large proportion of the population. This explains why this chapter not only proposes innovative methods for tackling concrete problems (such as defining a method for vaccinating people and defining how to verify this method) but also establishes a trade-off between the different dimensions of a vaccination strategy and protection strategy (the proposals aim to ensure an equilibrium between sanitary protection, citizen trust and economic resilience).

Figure 8.1 illustrates the authors' proposal to combine multiple dimensions to define an effective vaccination strategy and protection strategy against the Covid-19 pandemic.

To reach this objective, the authors propose a detailed literature review that classifies the existing literature into complementary dimensions. First, the review tackles different contributions that aim to optimize vaccination either by introducing some incentives such as a green pass or by using optimization models. Then, a focus is

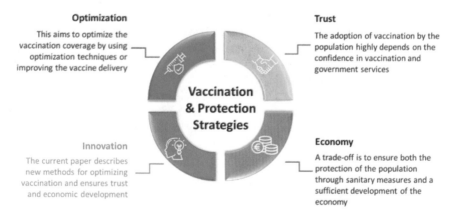

Fig. 8.1 The complementary dimensions considered to define innovations

laid on the importance of establishing trust between government services and citizens to ensure wide adoption of vaccination that favors herd immunity. Finally, an analysis of the research dedicated to establishing a trade-off between the sanitary and economic perspectives is provided to set the last basis before proposing innovations that contribute to the Covid-19 vaccination strategy. The innovation dimension will be addressed by the authors' proposals detailed in Sect. 8.3.

8.2 Literature Review

8.2.1 Optimizing Vaccination

The Israeli experience, which is ahead in terms of vaccination coverage compared to other countries, offers many relevant scientific observations [7]. First, it is crucial to pay attention to the hesitancy towards vaccination and to provide scientific information that mitigates anti-vax communication. Then, it is necessary to accompany some particular groups of the population that are more reluctant to vaccination for religious or cultural reasons. Moreover, it is difficult to maintain the vaccination level over time as the early adopters are the most confident about vaccination while the latter are the most reluctant. These conclusions are applicable in any other country.

A widespread strategy to optimize vaccination is to introduce a green card that corresponds to a quasi-mandatory vaccination. Whereas this approach is suitable to increase the immunization rate among the population, it also reduces the acceptance of a large part of the population [8], especially among people with lower education. Thus, it is of importance to offer education to the whole population. The green card attached to vaccination also raises some ethical concerns as it reduces the population's rights. The pass restricts the liberty of movement and the practice of particular jobs to preserve the population's safety. These restrictions should be limited and proportional to the objective pursued to fulfill ethical considerations [9].

The efficiency of a vaccination strategy also depends on the efficiency of vaccines and the acceptance of the population, which directly impacts its immunization. Assuming that anti-vaccine narratives spread around, it has shown that central nodes of the network are more exposed to these narratives to adopt them [10]. This may drastically reduce the efficiency of the vaccination strategy as indicated by the Susceptible Vaccinated Infected Recovered (SVIR) model proposed and increase the number of years lost due to the Covid-19 pandemic.

To fight Covid-19, many optimization strategies were detailed. A comprehensive review of prediction and control publications was proposed in Ref. [1]. This review proposes a detailed classification of optimization models depending on their perspective: (1) microscale versus macroscale, (2) early stages versus later stages, (3) aspects with direct versus indirect relationship to COVID-19, and (4) compartmentalized perspective. This paper offers optimization strategies that should be considered before initiating a new project to optimize vaccination.

A concrete vaccination strategy was proposed in Ref. [11] based on a simulation concerning Austria. To this end, the authors implemented decision-analytic modeling to compare multiple sequential prioritization rules that target different subgroups and strategies. Interestingly, the model compares strategies concerning slices of the population (above 65 years old, between 45 and 64, and between 15 and 44 years). They also take into account comorbidities that are more likely to increase the virus morbidity and the highest exposure to the virus among healthcare workers. With the model proposed in the paper, the results are optimal when the vaccination is prioritized with the elderly and vulnerable, middle-aged, healthcare workers, and younger individuals, respectively. However, as the vaccination reduces the contamination probability, the results could be different with a refined model where the contamination probability is higher for some groups of the population.

8.2.2   Building Trust

The introduction of vaccination incentives plays an important role in vaccination coverage as depicted in Ref. [12]. The increase in vaccination coverage in Israel correlates to the introduction of green pass incentives such as the possibility to access social, cultural, and sporting events for fully vaccinated or immune people. The authors suggest ensuring public confidence by adopting a sanitary approach that simultaneously ensures transparency of information and fair access to COVID-19 vaccination.

Increasing trust in vaccination certificates also contributes to increasing vaccination coverage. For example, a survey in Germany proved that paper-based vaccination certificates were favored over app-based solutions [13]. This result may be explained by the risk to reuse digital data for other purposes that would reduce privacy rights. However, the authors also observed that when vaccination certificates are mandatory for certain purposes, the type of certificate was of lower importance.

Another important battle against the pandemic is to identify people who have doubts about vaccination and convince them to protect themselves against the virus. An online survey performed in Australia in May 2020 was depicted in Ref. [14] and outlined that only 27% of the people were identified in the 'maybe' category. The factors favoring this position were an underestimation of the risks associated with the Covid-19 disease, low confidence in science, a high sensibility to misinformation and no former vaccination. Whereas this research is useful to outline the importance of focusing on the people that may change their minds, it does not provide a method for achieving this goal.

Currently, there is an effort to enhance the trust towards vaccination and reach herd immunity [15]. The objectives include taking advantage of social science to understand the elements that determine the population's willingness to vaccinate. The data obtained makes it possible to identify the psychological levers of the population and their associated behavior. The following observation was important: Motivation to vaccinate is both associated with rewards and incentives and the trust in authorities' communication. Both dimensions should be considered when it comes to optimizing the vaccination strategy.

Another interesting approach to optimizing vaccination is to identify and remove barriers that limit its adoption among the population. In Ref. [16], the author identifies both structural and attitudinal barriers. The structural barriers are defined as the limits of the health system and the country's organization to deliver the vaccine to people. The paper points out the importance of ensuring vaccine availability and the capability to associate a practitioner with every citizen. On the other hand, attitudinal barriers are the lack of confidence in a vaccine that uses a new technology (mRNA, messenger RiboNucleic Acid). The idea proposed in the paper is to build bipartisan endorsements and leverage social media platforms to promote vaccination.

8.2.3 Economy

Although the emergence of vaccine passports implies multiple limits in terms of ethics, it also offers a perspective to revive the economy and recover the pre-pandemic situation. This objective is of particular importance for countries that highly depend on tourism [17]. As a consequence, establishing vaccination passports contributes to a socioeconomic trade-off until herd immunity is achieved. Such a trade-off was already tackled in our previous chapter [18] with three combined dimensions (contamination rate, economic downturn and sociocultural impact). The vaccine passport is a new approach that may reduce the contamination while preserving the economic perspective provided that socio-cultural impacts are accepted (notably in terms of ethics).

An original approach to obtaining an equilibrium between sanitary and economic approaches is to question the priority of vaccination granted to older generations [19]. The authors observed that prioritizing the older part of the population saves more lives in the short term but slows down the economic recovery (even if young people tend to further spread the virus). The authors also introduced a utilitarian social welfare function that favors the vaccination of the elder ones, especially as new and more infectious variants appear. A refinement of the model would be to consider the impact on the education of children and a differentiation between first-line workers and back liners (i.e. teleworkers).

Another valuable method for defining the vaccination strategy is to implement a susceptible-exposed-infectious-recovered (SEIR) compartmental mathematical model [20]. This model relies on an adjustable relationship between lockdown intensity and economic loss (linear or non-linear functions). With the linear approach applied in Japan, it was proven that the better strategy is to vaccinate the younger generation, whereas in the second one the elder had to be vaccinated first. This research outlines the importance of defining the objective of a vaccination strategy to maximize its benefits.

The economic value of a COVID-19 vaccine was also tackled with a cost-effectiveness modeling using Markov chains [21]. The authors define the vaccine efficiency and its availability and estimate direct medical costs and deaths in the United States. The model also introduces an incremental cost per quality-adjusted life-year (QALY) gained versus no vaccine for different tiers of the population depending on their age and estimated risks. This research underlines the economic outcomes of vaccinating the population and the importance of ensuring a fast coverage of the population to reduce the overall morbidity of the pandemic.

To conclude this section, Table 8.1 summarizes the contributions and limits of the papers.

8.3   Examples of Innovations to Fight the Pandemic

8.3.1   The Neutralization Antibody Sustaining Method: A Mechanism to Maintain a High Level of Antibodies Among the Population

Several medical research publications showed that the immunity obtained through vaccination decreased after approximately 6 months with an immunity level below 50% after 3 months [22]. This requires setting up a vaccine booster strategy to ensure the long-term protection of the population. Indeed, stimulating the immune system every quarter ensures a continuous protection of citizens as the neutralization antibodies never fall under the minimal protection threshold. This evolution of immunity may be schematically illustrated in Fig. 8.2.

Thus, the authors proposed to refine the health pass by defining 3 groups of people:

- Group 1: fully vaccinated and immunized people
- Group 2: vaccinated people with incomplete immunization
- Group 3: unvaccinated people.

This categorization aims to reduce contamination risks by restricting access to the most exposed locations and infrastructures in Group 1 (i.e., locations where masks cannot be worn: restaurants, nightclubs, and other densely populated places). It also makes it possible to increase the protection required for under-protected groups by requiring safer protection (FFP2 masks for Group 2, FFP3 or higher for group 3). Figure 8.3 illustrates how the proposal may be implemented.

The protection mechanism aims to introduce a level between the vaccinated and unvaccinated levels. This offers a refined approach to reducing contamination and proposes a booster for vaccinated people with low protection. The test and secure stage functions as a barrier between the virus and people in indoor locations with contamination risks.

Optimally, the test mechanism relies on a fast evaluation of Covid protection to define whether vaccinated people need a booster to remain in the first group. However,

Table 8.1 Contributions and limits of the different paper cited in the literature review

Id	Contribution	Limits
[7]	Provides many recommendations to optimize vaccination through the analysis of social behavior	The paper is limited to the situation of Israel and only addresses the social dimension
[8]	Highlights the importance of education in the acceptance of Covid vaccination	This is a long-term proposal that does not contribute to rapidly resolve the pandemic
[9]	Points out that restrictions should be limited to fulfill ethical considerations	The paper does not propose!!! new measures to fight the pandemic
[10]	Uses the SVIR model to prove that central nodes are more likely convinced by anti-vaccine theories	The research does not provide new approaches to protecting the vaccination strategy from fake news
[11]	Details the case study of the vaccination strategy in Austria by slicing the population into!! age groups	The model is really specific to the case study and should be refined to consider various contamination probabilities
[12]	Outlines the importance of vaccination incentives and transparency	The paper does not propose new measures to fight the pandemic
[13]	Underlines that vaccination certificates tend to increase the protection coverage	The study is restricted to the case study of Germany and does not provide innovations
[14]	Highlights the necessity to convince undecided people to increase vaccination coverage	The paper does not propose new measures to fight the pandemic
[15]	Advocates for the usage of artificial intelligence to identify the levers of people towards vaccination	The paper does not indicate how to introduce incentives and rewards to optimize vaccination and reach herd immunity
[16]	Outlines the importance of suppressing barriers towards vaccination	The proposal is focused on social and organizational barriers
[17]	Highlights that vaccination passports contribute to a socioeconomic trade-off until herd immunity is achieved	The paper does not propose!! new measures to fight the pandemic
[18]	Advocates for the introduction of methods to set a trade-off between sanitary, social and economic dimensions	The proposal does not tackle the vaccination perspective
[19]	Questions the priority of vaccination granted to older generations to reach the optimal collective protection	The research is focused on the sanitary perspective to optimize vaccination
[20]	Implements a mathematical model to propose a strategy that ensures a trade-off	The proposal relies on the largely employed SEIR model
[21]	Proposes a cost-effectiveness modeling using Markov chains that offers high simulation capabilities	The proposal does not tackle the acceptability of the cost modeling proposed

Fig. 8.2 Illustration of the
Neutralization Antibody
evolution according to time

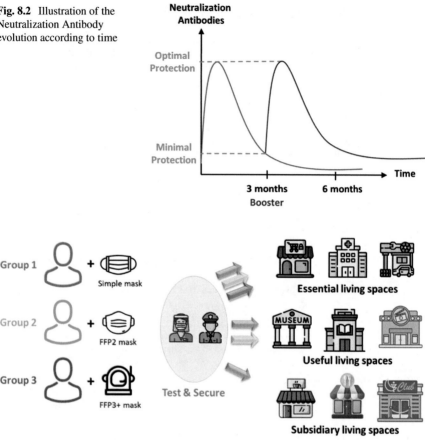

Fig. 8.3 Illustration of the proposed protection mechanism

a simple time verification based on the month delay may be sufficient to initiate the
process.

It is particularly important to underline that the system relies on a perfect protection of the verification squad that is directly exposed to the virus. The workers
assigned to this mission should have no comorbidity and should be easily replaced
by colleagues.

The applicability of the proposal is detailed in Table 8.2.

Table 8.2 Evaluation of the first proposal: The Neutralization Antibody Sustaining Method

Maturity of the proposed method	**High**. The method is largely organizational. This facilitates its immediate applicability and helps to fight the pandemic
Time frame for the expected availability of the proposed method	Immediately
Requirements to apply the proposed method	A legal protocol, a large number of Covid-19 tests, and a sufficient number of sanitary and security staff to implement the process
Expected efficiency of the proposed method	**Intermediary**. The separation of vaccinated people from others contributes to reducing contamination risks. It also reduces the probability that severe infections occur. However, it does not prevent every case of contamination

8.3.2 The Vaccination Fraud Detection Method: A Process to Identify Fraud

Whereas fraud detection is largely implemented in several domains of governmental intervention (fiscal fraud, unemployment insurance or social subsidies), there is limited research available concerning fraud by using a fake sanitary pass. Thus, the authors propose a process to identify fraud as illustrated in Fig. 8.4.

This method aims to identify potential fraud among the whole population. This requires gathering information on vaccination programs to be able to verify the vaccination status of citizens. Thus, a vaccination table offers the opportunity to identify practitioners that vaccinate a high number of patients (they may act honestly but a verification should be obligatory above a certain threshold).

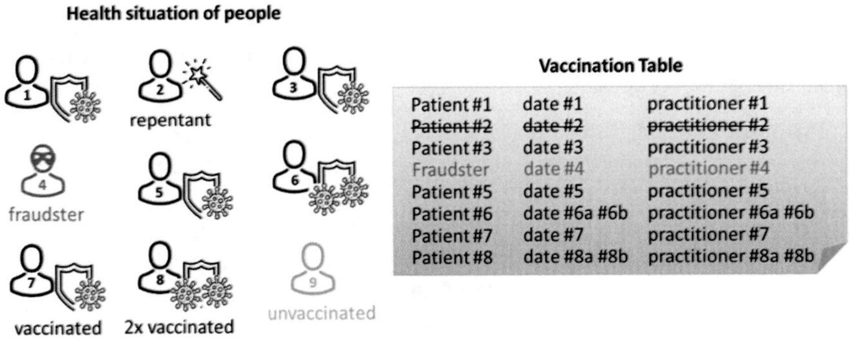

Fig. 8.4 Illustration of the proposed fraud detection mechanism

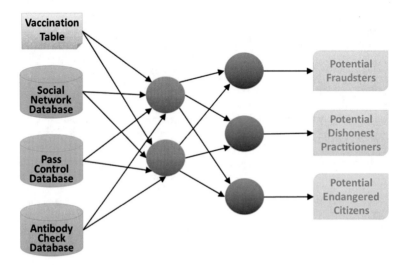

Fig. 8.5 Possible implementation of the detection mechanism

Then, the authors propose to implement an algorithm that combines other useful information by using artificial intelligence to identify potential fraudsters as illustrated in Fig. 8.5.

- **Social Network Database**: This database aims to collect information and aims to detect suspicious behavior of patients or practitioners;
- **Pass Control Database:** This database keeps track of each check of a vaccine pass and facilitates the detection of—areas or people that are not checked properly;
- **Antibody Check Database:** This database is dedicated to the analysis of antibody protection and offers opportunity to optimize the organization of vaccination boosts.

This proposal should be supplemented by the following principles with two orientations (prevention and detection). To avoid that citizens commit fraud, the authors make three initial proposals:

- **Proposal 1:** A practitioner may not vaccinate the same patient twice. This favors self-control as practitioners know that their patients will be checked by other practitioners.
- **Proposal 2:** A patient antibody charge may be checked at any time (especially when a vaccination pass must be shown). The risk to be checked reduces the probability that a patient commits a fraud.
- **Proposal 3:** A fraudster should pay a considerable fine or be sent to prison. This reduces the inclination to commit a fraud.

To ensure a high fraud detection rate, the authors make three additional proposals:

Table 8.3   Evaluation of the second proposal: The Vaccination Fraud Detection Method

Maturity of the proposed method	**Relatively high.** The method relies on modelization and neural networks. This requires some energy to collect data, combine it and generate outputs
Time frame for the expected availability of the proposed method	Rapid
Requirements to apply the proposed method	Collect data through a secured mechanism
Expected efficiency of the proposed method	**High.** The neural network makes it possible to identify multiple parts of the population. Then, government services should intervene to achieve concrete measures such as police investigations, arrests, precautionary measures for practitioners, or sanitary information and assistance to citizens

- **Proposal 4:** Ensure that each check of a vaccination pass is done together with an identity check.
- **Proposal 5:** Verify practitioners with suspicious vaccination statistics (highest rate of medical acts) or that have been punished by their medical institutions (for unappropriated Covid speech).
- **Proposal 6:** Screen social networks to detect anti-vaccination comments or behavior.

The applicability of the proposal is detailed in Table 8.3.

8.3.3   The School Block-Chain Detection Method: Recording the Interactions of School Children to Follow Back the Contamination Chain

Despite the effort of government services to improve the sanitary situation in schools, the contamination among children is difficult to anticipate as children often do not present symptoms of the disease and interactions occur at various moments. The proposal in this section is original and debatable as it relies upon recording each contact above a certain time threshold (example: 30 seconds).

The record should be destroyed after a very short time (at most 1 week). In this objective, each pupil must activate Near Field Communication (NFC) equipment to go to school and a 30-second contact triggers a signal to the nearest network equipment.

Then, contacts are stored anonymously using the block-chain technology (NFC#1, NFC#2, location, duration in seconds, level of risk) to make it possible to predict future contamination from the social interactions observed. Figure 8.6 illustrates the proposal.

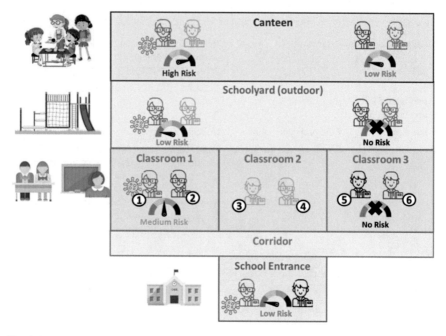

Fig. 8.6 Illustration of the proposed NFC recording mechanism

In the example above, several situations are discussed. The hypothesis is that pupils only wear a mask in indoor situations.

- Two children meet in an indoor situation for 1 min. This situation is recorded by the system with the following vector (#NFC1, #NFC5, entrance, 60, 0%). As the contamination of child #1 is not known at this moment, the contamination is estimated as no risk. However, if the child is screened as positive after two days (for example), then the level of risk could have risen to 20% (low risk).
- Two children work side-by-side in a classroom for 1 h. The system records the vector (#NFC1, #NFC2, classroom1, 1800, 0%). As soon as child#1 is screened positive, the level of risk is set to 50% (medium risk).
- Two children play together in the schoolyard for 5 min. The system records the vector (#NFC1, #NFC4, schoolyard, 300, 0%). As soon as child#1 is screened positive, the level of risk is set to 20% (low risk).
- Two children play and eat face to face in the canteen for 30 min. The system records the vector (#NFC1, #NFC4, canteen, 1800, 0%). As soon as child#1 is screened positive, the level of risk is set to 90% (high risk).

Table 8.4 illustrates a possible estimation of the contamination risk when exposed to a contaminated pupil. The table is based on the hypothesis that both children have a mask indoors and no mask outdoors.

The applicability of the proposal is detailed in Table 8.5.

Table 8.4 Evaluation of the contamination risk (example)

	Entrance	Classroom	Schoolyard	Canteen
Low	<5 min	/	<10 min	/
Medium-Less	≥5 min	<30 min	≥10 min	/
Medium–High	/	≥30 min	/	Side table
High	/	/	/	Same table

Table 8.5 Evaluation of the third proposal: The Vaccination Fraud Detection Method

Maturity of the proposed method	**Relatively high**. The method relies on NFC communication and a simple table of contamination. This requires some effort to gather data in the detection block chain
Time frame for the expected availability of the proposed method	Rapid
Requirements to apply the proposed method	Collect data through a secured mechanism Legal agreement
Expected efficiency of the proposed method	**Intermediary**. The system facilitates predictions concerning new contamination. However, it does not entirely prevent contamination as its functioning is based on quick action as soon as contamination is detected

8.3.4 The Sanitary Measures Optimization: Defining a Suitable Trade-off Between Measures Restricting the Individual Rights of Citizens and Protection of the Population Through Sanitary Measures

In this section, we propose a state transition-based model to characterize the various stages of a Covid infection and related state transition probabilities—while the exact transition probabilities are open for further debate, the overall approach is suitable to identify a reasonable compromise between maintaining the individual rights of citizens and protecting against a pandemic.

The proposed model enables us to model the contamination status using a Markov chain model. This implies that only the most recent health status affects what happens next. The principle is illustrated below. For example, a healthy person is in health status 1. This person has a probability to be infected and moves to status 2 that is expressed as $p_{1,2}$. Then, the status of the person may be 0 if the person obtains immunity, 3 if the virus causes symptoms, and 1 if the person recovers its health without immunity is indeed illustrated by Fig. 8.7.

It means that the random variable X_{t+1} depends upon X_t, but it does not depend upon X_{t-1}, or previous random variables.

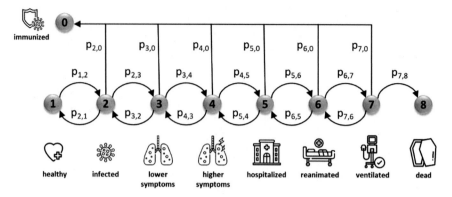

Fig. 8.7 Illustration of the transition model proposed

Define the following variables:

- $p_{i,j}$ the probability of transition from health status i to health status j (without initial vaccination)
- $\mathbb{P}$ is the transition matrix such as $p_{i,j} = \mathbb{P}\,(X_{t+1} = j \,|X_t = i)$.

$$P = \begin{bmatrix} p_{1,1} & \cdots & p_{1,N} \\ \vdots & \ddots & \vdots \\ p_{N,1} & \cdots & p_{N,N} \end{bmatrix}$$

The matrix values may be analyzed as follows:

- The diagonal values should be high since the health situation does not necessarily evolve every day.
- The transition towards an increase of the virus effects is represented by the values on the right of the diagonal $[p_{1,2};\ p_{2,3}\ ...;\ p_{N-1,N}]$
- The transition towards a decrease of the virus effects is represented by the values on the right of the diagonal $[p_{2,1};\ p_{3,2}\ ...;\ p_{N,N-1}]$
- The first column corresponds to obtaining immunity where people are considered "out of the process" (but a refinement could be to reintroduce them as the immunity decreases)
- The last column corresponds to death and people are definitely out of the process.

We define the transition table illustrated in Fig. 8.8.

To estimate the effect of a health protection measure, it is possible to introduce a new transition matrix so that the contamination is reduced. For instance, the obligation to wear a mask indoors may modify the matrix so that $p_{2,3}$ is reduced proportionally to mask adoption (if 80% of the population wears a mask then $p'_{2,3} = (1 - 80\%) \times p_{2,3}$). Another example is to introduce a vaccination pass that reduces $p_{3,4}$, $p_{4,5}$, and

1	0	0	0	0	0	0	0	0
0	0.8	0.2	0	0	0	0	0	0
0.05	0	0.75	0.2	0	0	0	0	0
0.03	0	0	0.77	0.2	0	0	0	0
0.02	0	0	0	0.78	0.2	0	0	0
0.01	0	0	0	0	0.79	0.2	0	0
0.001	0	0	0	0	0	0.799	0.2	0
0.0001	0	0	0	0	0	0	0.7999	0.2
0	0	0	0	0	0	0	0	1

Fig. 8.8 Illustration of the transition table proposed

$p_{5,6}$ similarly. The results obtained after 40 iterations (simple iterative multiplications of the initial vector by the transition matrix M) are illustrated in Fig. 8.9.

The simulation illustrated in Fig. 8.9 may be interpreted as follows:

- The red histogram indicates that the absence of masks and vaccination implies a fast convergence of the process and a high number of deaths.
- The green one highlights that the combination of masks and vaccines slows down the pandemic spread and also provides a high level of immunity with a low number of deaths.
- The blue one shows that the absence of vaccines implies a significant number of deaths even if the pandemic propagation is controlled.
- The yellow histogram outlines that the absence of masks accelerates the pandemic but the vaccine is fundamental to reducing the number of deaths. However, the situation could imply a very high level of deaths if the virus lethality was higher and/or the vaccine efficiency lower.

The proposal is completed with a method to optimize the constraint parameters. Indeed, the transition matrix should be optimized to take into account the new behavior of citizens (reduced efficiency of masks along time, increased transmission probability according to the virus variant, better efficiency of vaccines). This last step of the method is illustrated in Fig. 8.10.

The applicability of the proposal is detailed in Table 8.6

8.4 Conclusion and Perspectives

In this chapter, we have detailed and evaluated several measures to address the Covid pandemic—a corresponding comparison has indeed been introduced based on common criteria.

A first measure targets the objective to maintain a relevant level of antibodies and thus ensure resistance against Covid among the population. The second proposal

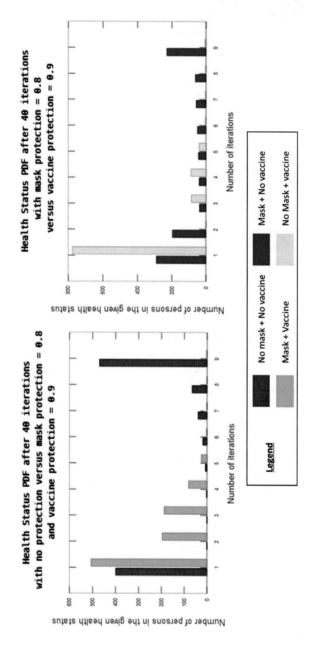

Fig. 8.9 Illustration of the results obtained by applying the Markov transitions

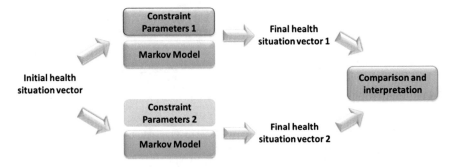

Fig. 8.10 Illustration of a comparison approach to optimizing parameters

Table 8.6 Evaluation of the fourth proposal: The Vaccination Fraud Detection Method

Maturity of the proposed method	**High**. The method is available for immediate application. It is open for further debate on how to improve estimates of state transition probabilities. Still, even approximate assumptions allow for an insightful simulation that can be exploited to define a suitable government strategy to address the pandemic
Time frame for the expected availability of the proposed method	Immediately
Requirements to apply the proposed method	A government-level implementation is required to implement the most suitable identified measures
Expected efficiency of the proposed method	**High**. Even if the model is run based on approximate sets of state transition parameters, it is expected that a suitable trade-off can be identified between the protection of the individual rights of a citizen and the introduction of measures to fight the pandemic

further extends those measures by introducing additional measures against vaccination fraud. A third approach suggests the usage of block-chain technology to track and characterize the spread of the pandemic among school kids.

Finally, a fourth method enables government agencies to model the expected evolution of a pandemic based on a state transition approach including related state transition probabilities.

We show how this method is appropriate to derive a suitable trade-off between protecting the individual rights of citizens and protecting the whole population. A combination of the proposed measures is expected to lead to the best strategy to fight the current pandemic. Furthermore, the underlying principles may be considered in a future pandemic or medical emergency.

Acknowledgements The authors would like to express their gratitude to Bettina Mertin for her valuable suggestions and careful review of the manuscript.

References

1. Jordan, E., Shin, D. E., Leekha, S., & Azarm, S. (2021). Optimization in the context of COVID-19 prediction and control: A literature review. *IEEE Access, 9,* 130072–130093. https://doi.org/10.1109/ACCESS.2021.3113812
2. Bertsimas, D., Digalakis Jr, V., Jacquillat, A., Li, M., & Previero, A. (2021). Where to locate COVID-19 mass vaccination facilities? *Naval Research Logistics (NRL).* https://doi.org/10.1002/nav.22007
3. Parino, F., Zino, L., Calafiore, G. C., & Rizzo, A. (2021). A model predictive control approach to optimally devise a two-dose vaccination rollout: A case study on COVID-19 in Italy. *International Journal of Robust and Nonlinear Control.* https://doi.org/10.1002/rnc.5728 (Advance online publication).
4. Jadidi, M., Jamshidiha, S., Masroori, I., Moslemi, P., Mohammadi, A., & Pourahmadi, V. (2021). A two-step vaccination technique to limit COVID-19 spread using mobile data. *Sustainable Cities and Society, 70,* 102886. https://doi.org/10.1016/j.scs.2021.102886
5. Bhardwaj, A., Ou, H. C., Chen, H., Jabbari, S., Tambe, M., Panicker, R., & Raval, A. (2020). Robust lock-down optimization for COVID-19 policy guidance. In *AAAI Fall Symposium.* https://projects.iq.harvard.edu/files/teamcore/files/robust_lock-down_optimization_for_covid-19_policy_guidance.pdf
6. Meehan, M. T., Cocks, D. G., Caldwell, J. M., Trauer, J. M., Adekunle, A. I., Ragonnet, R. R., & McBryde, E. S. (2020). *Age-Targeted Dose Allocation can Halve COVID-19 Vaccine Requirements.* medRxiv 2020.10.08.20208108; https://doi.org/10.1101/2020.10.08.20208108
7. Rosen, B., Waitzberg, R., Israeli, A., et al. (2021). Addressing vaccine hesitancy and access barriers to achieve persistent progress in Israel's COVID-19 vaccination program. *Israel Journal Health Policy Research, 10,* 43. https://doi.org/10.1186/s13584-021-00481-x
8. Gallè, F., Sabella, E. A., Roma, P., Da Molin, G., Diella, G., Montagna, M. T., Ferracuti, S., Liguori, G., Orsi, G. B., & Napoli, C. (2021). Acceptance of COVID-19 vaccination in the elderly: A cross-sectional study in Southern Italy. *Vaccines, 9*(11), 1222. https://doi.org/10.3390/vaccines9111222
9. Kamin-Friedman, S., & Peled Raz, M. (2021). Lessons from Israel's COVID-19 green pass program. *Israel Journal Health Policy Research, 10,* 61. https://doi.org/10.1186/s13584-021-00496-4
10. Prieto Curiel, R., & González Ramírez, H. (2021). Vaccination strategies against COVID-19 and the diffusion of anti-vaccination views. *Science and Reports, 11,* 6626. https://doi.org/10.1038/s41598-021-85555-1
11. Jahn, B., Sroczynski, G., Bicher, M., Rippinger, C., Mühlberger, N., Santamaria, J., Urach, C., Schomaker, M., Stojkov, I., Schmid, D., Weiss, G., Wiedermann, U., Redlberger-Fritz, M., Druml, C., Kretzschmar, M., Paulke-Korinek, M., Ostermann, H., Czasch, C., Endel, G., … Siebert, U. (2021). Targeted COVID-19 vaccination (TAV-COVID) considering limited vaccination capacities—An agent-based modeling evaluation. *Vaccines, 9*(5), 434. https://doi.org/10.3390/vaccines9050434
12. Saban, M., Myers, V., Ben Shetrit, S., & Wilf-Miron, R. (2021). Issues surrounding incentives and penalties for COVID-19 vaccination: The Israeli experience. *Preventive medicine, 153,* 106763. https://doi.org/10.1016/j.ypmed.2021.106763
13. Kowalewski, M., Herbert, F., Schnitzler, T., & Dürmuth, M. (2022). Proof-of-Vax: Studying user preferences and perception of Covid vaccination certificates. *Proceedings on Privacy Enhancing Technologies, 2022*(1), 317–338. https://doi.org/10.2478/popets-2022-0016

14. Attwell, K., Lake, J., Sneddon, J., Gerrans, P., Blyth, C., et al. (2021). Converting the maybes: Crucial for a successful COVID-19 vaccination strategy. *PLoS ONE, 16*(1), e0245907. https://doi.org/10.1371/journal.pone.0245907
15. Van Oost, P., Yzerbyt, V., Schmitz, M., Vansteenkiste, M., Luminet, O., Morbée, S., Van den Bergh, O., Waterschoot, J., & Klein, O. (2022). The relation between conspiracism, government trust, and COVID-19 vaccination intentions: The key role of motivation. *Social Science and Medicine, 301*, 114926. ISSN 0277-9536. https://doi.org/10.1016/j.socscimed.2022.114926
16. Fisk, R. J. (2021). Barriers to vaccination for coronavirus disease 2019 (COVID-19) control: Experience from the United States. *Global Health Journal, 5*(1), 51–55. ISSN 2414-6447. https://doi.org/10.1016/j.glohj.2021.02.005
17. Sharun, K., Tiwari, R., Dhama, K., Rabaan, A. A., & Alhumaid, S. (2021). COVID-19 vaccination passport: Prospects, scientific feasibility, and ethical concerns. *Human Vaccines and Immunotherapeutics.* https://doi.org/10.1080/21645515.2021.1953350
18. Gaie, C., & Mueck, M. (2022). An artificial intelligence framework to ensure a trade-off between sanitary and economic perspectives during the COVID-19 pandemic. In *Deep Learning for Medical Applications with Unique Data* (pp. 197–217). Academic Press. ISBN 9780128241455. https://doi.org/10.1016/B978-0-12-824145-5.00008-3
19. Glover, A., Heathcote, J., Krueger, D., & Ríos-Rull, J.-V. (2021). *Optimal Age-Based Vaccination and Economic Mitigation Policies for the Second Phase of the Covid-19 Pandemic.* https://cpb-us-w2.wpmucdn.com/web.sas.upenn.edu/dist/9/544/files/2021/11/vax.pdf
20. Sunohara, S., Asakura, T., Kimura, T., Ozawa, S., Oshima, S., et al. (2021). Effective vaccine allocation strategies, balancing economy with infection control against COVID-19 in Japan. *PLoS ONE, 16*(9), e0257107. https://doi.org/10.1371/journal.pone.0257107
21. Kohli, M., Maschio, M., Becker, D., Weinstein, M. C. (2021). The potential public health and economic value of a hypothetical COVID-19 vaccine in the United States: Use of cost-effectiveness modeling to inform vaccination prioritization. *Vaccine, 39*(7), 1157–1164. ISSN 0264-410X. https://doi.org/10.1016/j.vaccine.2020.12.078
22. Levin, E. G., Lustig, Y., Cohen, C., Fluss, R., Indenbaum, V., Amit, S., Doolman, R., Asraf, K., Mendelson, E., Ziv, A., Rubin, C., Freedman, L., Kreiss, Y., & Regev-Yochay, G. (2021). Waning immune humoral response to BNT162b2 Covid-19 vaccine over 6 months. *The New England Journal of Medicine, 385*(24), e84. https://doi.org/10.1056/NEJMoa2114583

Chapter 9
From Paper to Digital: e-Government's Evolution and Pitfalls in Brazil

Fabrício Ramos Neves⊙ and **Polyana Batista da Silva**⊙

Abstract This chapter provides an overview of the e-government processes in Brazil. We highlight some recent initiatives, emphasizing development, challenges, the relationship with other levels of government, and best practices while providing a case for the country. Despite Brazil's recent political, economic, and financial crises, e-government is an ongoing process. However, it can be discontinued if actions encouraging digital tool use among citizens and civil servants are wrongly set up. We argue that digital transformation requires more than just data and algorithms and is enmeshed in social relationships. Regulation serves as a foundation, but e-government is a dynamic process, much more than laws are required. By illustrating the evolution and highlighting some examples in this chapter, we hope to inspire practitioners, legislators, and policymakers to consider better public policies for participation and digital transformation to improve e-government processes.

Keywords e-Government · Digital Government · Digital transformation

9.1 Introduction

The role of information and communication technologies in the most diverse aspects of our daily lives is becoming increasingly evident in the current context, in which all countries are dealing with the COVID-19 pandemic and its social and economic consequences. As digital technologies pervade all aspects of society and the economy,

F. R. Neves
Federal Institute of Education, Science and Technology Baiano, Guanambi, Brazil

P. B. da Silva
School of Economics, Business Administration and Accounting at Ribeirão Preto - University of São Paulo, Ribeirão Preto, Brazil
e-mail: polyanasilva@usp.br

F. R. Neves (✉)
School of Arts, Sciences and Humanities - University of São Paulo, São Paulo, Brazil
e-mail: fabricioneves@alumni.usp.br

© The Author(s), under exclusive license to Springer Nature Switzerland AG 2023   193
C. Gaie and M. Mehta (eds.), *Recent Advances in Data and Algorithms for e-Government*,
Artificial Intelligence-Enhanced Software and Systems Engineering 5,
https://doi.org/10.1007/978-3-031-22408-9_9

it is critical that all actors—in the private or public realm—who may be affected by their use are aware of their potential consequences.

This complex scenario has required governments to make rapid progress in adopting information and communication technologies in a variety of sectors, including business, education, commerce, and health care, among others. Faced with the digital transformation we are witnessing—an economy driven by data and applications based on artificial intelligence—there is a global race to lead critical aspects of the development of associated technologies in a combination of intellectual and financial efforts that will give developing countries some advantages. AI-based technologies could play a critical role in stimulating socio-economic development in emerging nations, whether to increase such comparative advantages or to improve the quality and efficiency of services provided by government agencies to the population.

Understanding algorithms' positive and negative social, political, and economic impacts is a research challenge and a relevant topic for governance mechanisms and public policies. In addition to this challenge, the diversity of data and algorithms in use and their complexity are rapidly growing [1].

Data and algorithms have been used in many ways. It can help manage work and workers [2, 3]; healthcare [4]; human resources [5]; traffic in large urban centers [6]; financial decisions [7]; increase the efficiency of government services [8], and fight against corruption [9].

On the other hand, data and algorithms designed to filter, select, and display the vast amount of online information can induce bias and create groups in society, fueling racism, prejudice, and discrimination [2, 10].

The way for governments to have control and governance over how algorithms should operate and how their data can be used has been through the establishment of norms and rules, and they must make an effort to understand the daily interactions with human beings to create a digital governance environment capable of guaranteeing their use for public services.

Although connected with the application of information technology in the public sector, the concept of electronic government has a larger scope. It is linked to the modernization of public administration through the use of information and communication technology, as well as the improvement of the efficiency of government technical and management operations [11].

When analyzing the widespread use of internet resources and the evolution of the information society, various aspects need a more consistent evaluation of e-government directions. This is where the focus of this chapter lies. We use the Brazilian context as a case study to provide an overview of the e-government process, contrasting federal and local government initiatives to demonstrate the country's e-government development, challenges, and best practices, which can provide some insights into other countries.

9.2 State Reform and the Emergence of Electronic Government in the Brazilian Government

Much of the research on e-government in Brazil dates the emergence of e-government back to the year 2000 [12, 13]. However, we understand electronic government as a collection of modernizing actions related to public administration that gained prominence in the late 1990s [14]. In 1996, the country was the first in the Americas to deploy electronic voting [15]. Since then, it has moved to an Internet-based system.

Measures to modernize the public sector in Brazil have been in place since the 1970s, but with the fiscal crisis of the 1980s, state intervention became known as a type of public management reform. During this period (1970–1990) of emphasis on internal management [16], governments began implementing structures to plan and coordinate computing operations. A significant amount of data from bureaucratic management was generated, which was essential to the country's eventual adoption of electronic government. On the other hand, their initiatives were weak and unconnected to government plans and public policy.

One of the reasons is that investments in technology and human resources for IT were jeopardized at the end of the military regime (in early 1985), when the state's fiscal crisis was at its worst, weakening the public information technology structure. As the systems were designed for operational control rather than strategic purposes, investments in the area were not prioritized. Parallel to this, as part of the transition to democracy, the 1988 Brazilian Constitution decentralized some public policy responsibilities from the federal government to the States and Municipalities, resulting in a disconnection between government levels to accomplish the government plan of public management reform. Municipalities struggled for some time as States and the federal government provided municipal services for a while [17].

State reform and public sector modernization initiatives accelerated due to the fiscal crisis during the 1980s and the exhaustion of bureaucratic management and government intervention. In the 1990s, New Public Management concepts heavily influenced the Brazilian administrative and public management reform movements [17].

These reforms influenced the development of electronic government through the introduction of new management tools that relied on technical advancements in information and communication technologies (e.g., planning, budget, organizational structure, human resources, process management, and government procurement) characterized by innovations in both technology-related management and technological development [12]. The use of technological advancements is a key aspect of the modernization of public administration. The electronic government is one of the major reforming projects in Brazilian public administration, which overtime gave rise to four types of e-government services: Government-to-Business (G2B); Government-to-Citizen (G2C); Government-to-Employee (G2E) and Government-to-Government (G2G).

Nonetheless, issues such as (i) the balance between transparency and confidentiality of citizens' data; (ii) issues related to the integration of government data;

(iii) technological compatibility (interoperability); and (iv) budgetary continuity and bureaucratic rigidity should be kept in mind as major challenges and critical factors in the execution of an e-government project [18], pointing out that the proper treatment of technological, organizational, legal, and political issues is critical to success.

9.2.1 Access to Public Services and Data Information (1990–1998)

After the nation's re-democratization effort (around the end of the 1980s), it started to take steps to organize information derived from public data. For instance, in January 1991, the National Policy on Public and Private Archives (Law 8159/1991) was enacted, which stipulated the State's obligation to safeguard, protect, and manage documents and archives of a public nature and/or interest, which must be viewed as fundamental instruments to support administration, culture, and as a historical record of the nation's evidence and information. The law defined the concepts of public and private archives and dealt with the organization and administration of public archival institutions.

Pressure for transformation was brought on by the implementation of services for citizens. For instance, following the country's re-democratization, the Brazilian electoral system started to be automated. This process aimed to reduce the number of frauds and speed up the counting process. The Electronic Voting Collector was created in 1995 as part of a project to develop an electronic voting machine. It was used for the first time in 57 Brazilian cities in the municipal election of 1996, and by 2000, electronic voting machines had been used in all Brazilian municipalities. The so-called electronic democracy arose due to information and communication technologies in public management as a Government-to-Citizen groundbreaking experience in the country [19].

Despite the development of some electronic services, such as electoral services, internet access in the country was still in its early stages in the early 1990s. The Internet became widely available in Brazil in 1995, following what was happening worldwide [20]. The Brazilian Internet Steering Committee was established in 1995 but only fully implemented in 2003 (Decree 4829/2003). Its purpose is to develop strategic recommendations for the country's use and expansion of the Internet. It also evaluates network security measures, conducts surveys, and answers to legislation governing domain name and Internet Protocol (IP) address registration.

Another fundamental factor for the growth of the internet in the country was the privatization of telecommunications, which began in 1998 and gave various companies access to infrastructure and millions of potential clients. A tax incentive program encouraged people to buy computers, lowering the cost of the equipment, as with tablets and smartphones, allowing more Brazilians to connect to the internet network [20]. These actions paved the way for internet-based service delivery and e-government development.

9.2.2 Internet-Based Service Delivery (1999–2018)

In the early 2000s, the spread of the internet and the service delivery movement, which had as its main goal the pursuit of excellence and citizen-oriented services, combined with ICTs, enabled Brazilian governments to provide electronic public services to the population [21]. Technological improvements facilitated the expansion of electronic services, public communication channels' availability, and government objectives' disclosure. Since then, citizens have had continuous new ways to interact with the government, allowing them to express their opinions on government policies, monitor and control their actions, and express their needs, even though much work remains to be done [11].

The National Electronic Government Program was established in 2000 when an Interministerial Working Group was formed to examine and propose rules, standards, and norms linked to new electronic modes of engagement. The initiative aspired to become a single hub for societal services and information (the first version of the environment went live on January 25, 1999). Previously, the services were gathered under the Federal Government Services Portal, but they are now in a single platform called 'Gov.br'. Around a third of services were digital at the time. Currently, 60% of the 3300 services available are already available.

Meanwhile, in 2004, the Brazilian Office of the Comptroller General (CGU) launched the Transparency Portal, a platform aimed at improving fiscal transparency in the Brazilian Federal Government through open government budget data. The Transparency Portal was developed in collaboration with the Federal Data Processing Service (SERPRO) to enhance transparency and provide a platform that encourages citizen participation. The project is recognized as one of the most significant e-government projects in fiscal control. Similar projects have been built after Brazil's Transparency Portal by local governments and other Latin American countries.

Despite its positive reaction and relative success, the Transparency Portal continues to face some obstacles to further growth, stemming from several legacy issues that have existed since the National Policy on Public and Private Archives was established in the early 1990s, regarding the methods used historically in Brazil for gathering, storing, and disseminating information. For instance, one challenge has been unifying data from various government agencies due to the complex interrelationships of information technology, organizational culture, and normative regulations. This is true for legacy data and systems, which may have incompatible file formats. Their classification as 'legacy' suggests that nobody is keeping them updated, and they may suffer from compatibility issues with more recent hardware or software. Dealing with such diverse data formats and systems not only slows down digital transformation but also increases the risk of error introduction. At that time, interoperability and data standards from many government entities were still challenging.

To overcome these problems, the e-PING framework (Electronic Government Interoperability Standards) was initially institutionalized in 2005. This architecture aims to establish a minimal set of assumptions, rules, and technological requirements that govern the use of ICT in the interoperability of electronic government services, setting the stage for interaction with other powers, spheres of authority, and society at large. Citizens are favored by its adoption when, for example, requesting benefits from the government, such as Bolsa Familia (a social welfare program), which requires the integration of different data that may be spread across different levels of government, including those related to proof of income, school records, and vaccination card.

Brazilian Civil Rights Framework for the Internet (Law 12,965/2014) was approved by Congress and represents a turning point in international law by ensuring the civil rights of internet users are protected. They were designed as a piece of legislation that might uphold the principles behind the promotion of freedom and rights on the web. Distancing itself from an oppressive regulation, Brazil provided one of the most emblematic models to inspire international discussions on Internet regulation that upholds human rights as its guiding principle, retaining its central tenet to prevent an early expiration of its legal tools [22].

Several initiatives from this period enabled the development of algorithms and data structures to create e-government initiatives that leverage infrastructure and the collection of government data to provide transparency and e-services. However, many of the services that began to be provided by the internet were split among multiple organizations. Citizens had to complete multiple registrations, frequently with the same information and creating access passwords that were easily forgotten. The solution was the adoption of e-government interoperability standards in 2018 and introducing of the single 'Gov.br' portal in 2019. This has enabled the sharing and reuse of data among multiple government departments, as well as a reform of the state's services.

In Appendix A, we categorized the cited references based on their contribution to the main area of research.

9.2.3 Brazilian Data Protection Law (LGPD), Citizens' Data and Government

Another significant step in Brazil is the adoption of the Brazilian General Data Protection Law (Law 13,709/18). Despite its youth, the Law's foundations are comparable to those of the General Data Protection Regulation (GDPR), a law passed in 2018 to protect European citizens' data.

The obligation to protect Brazilian individuals' data does not apply only to private firms; it also applies to public organizations. As demonstrated by the examples in this chapter, the public sector processes citizens' data not only for the development and implementation of public policies but also for the delivery of the most diverse public services, such as enrollment of their children in public schools, monitoring of public health policies, and retirement programs.

The Brazilian public administration increasingly anticipated employing information and communication technology and data processing techniques as vital tools for public management and citizen involvement. It should be mentioned that the government's shared use of citizens' data must be carried out only for public purposes to execute legal competencies or complete the legal attributions of the public service. Thus, 'Gov.br', which aggregates a significant portion of public services (so far), is the government's database that exchanges citizen data with other public service platforms, such as the Digital Identification System, the vaccination system (ConecteSUS), and others.

In cases of a violation of the law as a result of personal data processing by public bodies, the Data Protection National Authority (The federal public administration body responsible for ensuring personal data protection accountable for implementing and monitoring compliance in Brazil) may send a report with appropriate measures to put an end to the violation, in addition to requesting public agents to publish impact reports on personal data protection and securing personal data.

9.2.4   Digitization and Platformization (2019–Nowadays)

Government platforms have been formed due to the connection and integration of information systems and databases developed for governmental operations and services in the 1990s and 2000s to promote information cooperation. Using artificial intelligence-based algorithms and a governance platform, this approach has also made it possible for e-government functions to be automated [8, 23–25]. Many services provided by the federal government are now accessible through platforms and applications (apps).

One example is the implementation of the 'Gov.br' platform, a digital ecosystem of engagement and communication between citizens and the Brazilian federal government. All federal government agencies and enterprises started to migrate their electronic addresses to this ecosystem in 2020. However, digitization is progressing slowly, and the goal is to provide 100 percent of the Federal government's services online (more than 3 thousand) by the end of 2022. To access numerous public services such as health care, education, social security, notary services, and others, users are identified digitally (through facial biometrics or an individual password).

A Government-to-Government (G2G) example is SICONFI, a project in development since 2014 with complete application in 2020. SICONFI is a web platform that uses XBRL (Extensible Business Reporting Language) as the standard for fiscal and financial data transmission from 5570 municipalities and 27 states to standardize consolidation procedures and improve the quality and reliability of financial and fiscal report statistics. It is also meant to boost openness in public administration, upgrade internal accounting procedures, support public sector managers in decision-making, and enable citizens to exercise citizenship through social control. The platform is under the supervision of the National Treasury, which has produced an Open Data Services—Application Programming Interface (API). Using the API, the user (citizen or otherwise) can obtain small fractions to huge volumes of data from all of the information entered by subnational entities in SICONFI. It is also a G2G service that allows transactions between the central/national and governments of other levels.

An example of Government-to-Employee (G2E) services is the digital platform 'Sougov.br', introduced in 2021 as part of a significant digital transformation effort to provide larger and better connectivity between current, retired, and pensioner employees of the Federal Public Administration. 'Sougov.br' is available in the app and web versions to be accessed via cell phone or computer. It provides a variety of Human Resources management services, including sending a medical certificate, requesting transportation assistance, maternity, adopting, and paternity leaves, and digital proof of life, in addition to access to the paycheck. Government-to-employee services include specific services solely available to civil servants, like human resource training and development, via digitizing over 45 services and developing over 20 new applications that will automate much of the services to public servants. The platform is expected to provide consumers with roughly 70 services by 2023.

We show in Fig. 9.1 a timeline with some key facts about the Brazilian digitization process.

In 2022, some initiatives will be put to the test. For example, the 'Gov.br' project was expanded with the development of 'Cidades.gov.br,' a standardized environment capable of delivering information on municipal services. Local governments will be able to rely on the 'Gov.br' platform to expand the range of services available to citizens. For the time being, it provides services such as tree pruning requests or notices, waste collection, road pothole alarms, and school enrollment space reservations in public schools. The interoperability of the 'Gov.br' and 'Cidades.gov.br' ecosystems can provide quick access to federal, local, and state government services. Cities that already offer some form of internet access will be able to join the 'Cidades.gov.br' platform. The platform was created to make the work of the public manager easier by providing a configurable interface and simple content management tools, allowing data reuse and improving services in health, education, and so on.

The Covid-19 pandemic prompted telemedicine on an experimental and temporary basis in 2020 to handle the situation in Brazil. The Brazilian government's 'Coronavirus SUS' app is an example. The application's concept was to map potential contamination locations and provide knowledge to identify possible virus contractions and how to prevent them [26]. Patients were routed to the local healthcare facility for testing if diagnostic replies supported by a questionnaire directly in the app suggested a possible infection, boosting the efficiency and effectiveness of the traditional healthcare setting. Combining preventive, screening, and information in one app has evolved into a telemedicine tool, decreasing the demands on Brazil's public healthcare system and, as a result, aiding in the prevention of the healthcare system's collapse. In addition, the application enabled the issuance of vaccination certificates to confirm immunity against the virus in conjunction with data from local governments. Despite its use, the Federal Council of Medicine did not consider telemedicine legal and was not regulated until 2022.

After the experience with telemedicine apps, the Brazilian federal government launched the Basic Digital Health Unit (UBS Digital) project in June 2022, a public

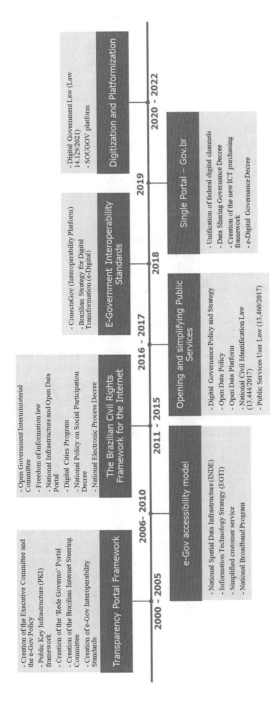

Fig. 9.1 Key facts of the Brazilian digitization process in a timeline

service offered in 323 Brazilian municipalities to increase access to online care in places outside of major urban areas. The municipalities involved in the trial project will be equipped with a 5G connection to facilitate consultation and monitoring operations such as remote examinations. The idea is that time and resources will be saved while health services would be delivered more effectively.

Another future development is the National Civil Identification program. Established in 2017 (Law 13,444/2017), the initiative aims to consolidate the issuance of identification credentials in Brazil. Before this initiative, each state distributed an identification card to each person. So, suppose a citizen is required (or desired) to issue a new identification certificate in one of the 27 different federation states. In that case, one might have up to 27 separate ID numbers, which opens possibilities for fraud, for example. With centralized issuance, digital identity unites citizen data, increasing citizen security and simplifying identity data management for state and local governments.

In theory, the database of civil registration offices on the national territory would be interoperable with the biometric database of the Brazilian Superior Electoral Court (with more than 120 million citizens registered in an electronic file with photos, signatures, and fingerprints). The National Identity Document is part of the National Civil Identification initiative, which aims to integrate a national citizen identification system across the country.

Implementing digital ID is the duty of the Public Security Secretariats via their Identification Institutes and will allow governments to act interoperable, i.e., sharing information among themselves. Blockchain technology strengthens the security of the digital ID, and each registered citizen becomes part of the cryptographic key generated by algorithms. This means that citizens' biographical and biometric data, such as facial recognition or fingerprints, are included in this security technology. According to the Brazilian Superior Electoral Court, the National Identity Document will be a keyway to identifying citizens in their interactions with society and governmental and private agencies and entities. It is expected to be available to the entire population by February 2023.

The Federal government is developing the Real Digital (Central Bank Digital Currency) to establish a digital currency. This is not a cryptocurrency but a digital counterpart of the Real (the country's present currency), which has the same backing as the actual money. One criterion for adoption in 2024 is interoperability between the digital version and the already available payment methods. The goal is for the end-user to have a virtual wallet in possession of a Central Bank-approved agent. Some steps are being taken to help the Brazilian digital currency develop. For example, the Central Bank established a study group in August 2020 to carry out studies on the issuance of a digital currency by the institution; another initiative is the Real Digital Challenge organized by the Financial and Technological Innovations Laboratory, the LIFT Challenge Real Digital, which aims to assess the use cases and technological feasibility of digital currency. The LIFT Challenge Real Digital team hopes to receive contributions from society and form a project plan. Pilot studies for payments, purchases, transactions, investments, use in smart contracts, international fees, and the internet of things are among the possible services. The potential reduction in the issuance of paper money, increased accessibility of banking services, ease of taxation, and extension of money monitoring, which would prevent money laundering crimes, are among the reasons for the deployment of digital currency in Brazil and other countries. Data sharing in developing and using a digital currency implies interoperability between banks, government organizations, and private firms.

9.3   Challenges in e-government Services

The country faces some challenges in implementing and carrying out e-government in Brazil. According to data from the OECD [27], Brazil needs to improve its broadband implementation. According to the data, Brazil's digitization budgets should be managed by a single management structure. Another issue confronting Brazil is the digital divide [10], which creates a partition between the wealthy and the poor and those who have and do not have access to the internet.

According to the OECD [27] report, Brazil's National Digital Signature and Identification System needs to be implemented as soon as possible. According to the report, legislation must be approved for the nation's society to go digital. It serves as a reminder of the all-government strategy. This specifically relates to electronic government, where all essential parties must cooperate to accomplish a common objective. It should not be forgotten that Brazil already has a government website (www.gov.br) where the public may access a variety of e-services as well as helpful information and data.

All nations must consider how safe citizens' personal information can be used. According to the Brazilian Digital Government Review [28], government officials must be aware of the value of technology as a tool for carrying out their everyday tasks. This covers e-government and how it affects how government workers conduct their daily business. The government must ensure that its regulations reflect the latest technological developments. The COVID-19 pandemic had a detrimental effect on economic factors, which the government must also consider, leading to significant adjustments in several areas.

The interaction between the federal, state, and local governments is a factor that needs to be considered. The smooth implementation of e-government across the nation depends on the efficiency of intergovernmental connections. The states' historical IT management problems, such as a lack of alignment with governmental policy and departmental needs, have not yet been fully overcome despite the deployment of e-government. For example, at the beginning of the pandemic, people who lived in remote regions of the country without access to the Internet or mobile phones found difficulty accessing the registration platform for the emergency assistance program. Homeless people also encountered difficulties, adding to the aggravating factor that many homeless and remote residents did not even have identification documents, which made them invisible to the federal government's identification systems.

Some restrictions connected with unstable technology infrastructure, for example, can hinder public organizations' connection with residents; data sharing and processing; and online access to information or services. Additionally, the use of cutting-edge digital technologies like Big Data, AI, the Internet of Things, and cloud computing depends on high-speed Internet access to ensure the gathering, processing, and analysis of data and information that will be used to create and implement evidence-based public policies.

In terms of ICT infrastructure in government agencies, the results of the ICT Electronic Government 2019 survey [29] confirmed optical fiber connections as the primary mode of Internet access in both federal and state (94%) and municipal agencies (73%). However, there are still some hurdles to expanding access via optical fiber, particularly in the North (54%) and Central-West (58%) regions, where just over half of municipalities have this sort of connection as one of the modes of internet network access [21]. Outsourcing portals in small local governments remains challenging [30].

Significant infrastructure investment is still necessary to deliver quality Internet connections in rural areas. Because just 10% of rural residents have an internet connection, they are disproportionately left out [29]. There have been reports of both paucities, lack of sites, and problems with the signal quality available; it is also critical to be able to execute digital inclusion initiatives in low-income neighborhoods of major cities, notably considering the job market's expectations as a consequence of the current increased use of electronic environments.

9.4 Concluding Ideas

The adoption of digital technologies to improve access to information and public services depends on creating technology solutions tailored to the demands of individuals, businesses, and other sectors of society. As a result, e-government or digital government strategies should encourage the development and implementation of information and communication technologies in the public sector to simplify and improve performance, promote innovation, and create value for businesses and society while ensuring long-term development and reducing inequalities.

In crises and emergencies, such as the COVID-19 pandemic, ensuring the availability of public services based on the intensive use of digital technologies is critical because such situations require public entities to either increase demand for already available services or create new policies to deal with the adversities imposed by the new context. For example, we witnessed situations indicating the necessity to strengthen health systems and provide aid to the growth of guidance to the people on their actions and tactics to reduce the damage caused by disease transmission.

We examined recent developments in digital transformation in several fields in Brazil. The first step toward successful digital transformation is developing a clear and coherent digital strategy. While regulations are essential, much more must happen for e-government to be a game-changer. For example, as stated in the chapter, the availability of high-quality fixed and mobile communication services is critical for Brazil's digital transition. In this regard, governments should take additional steps to promote high-quality broadband access, such as better coordination between the federal, state, and municipal levels, especially in underserved rural and remote areas, through investments in infrastructure, new technologies, and community networks. Along with roads and water infrastructure, broadband expansion is gaining prominence with a growing demand for modern technology.

While Internet access should remain a priority, Brazil should implement a broader set of policies to address the digital divide, such as developing services and content to meet the needs of low-income, low-educated, and elderly people, which should be coordinated at all levels of government. This could pave the way for citizens to use information and communication technologies more effectively in their civic engagement.

Continue to improve interoperability among public administration systems by adopting open data and interface standards; reinforce the establishment of the digital identity framework—National Digital Signature and Identification System in Brazil (currently under implementation); promote public and private innovation, as well as improve security and trust in a digital environment; and, finally, digital strategy should be supported by adequate resources. We believe that a similar approach could be helpful as a benchmark for other developing countries with a recent history of implementing e-government.

The Brazilian government took decades to build a regulatory framework at today's level, which must be an ongoing process. Digital transformation is already being handled through federal government initiatives to create rules and policies for the country's entry into the digital economy, such as those detailed in this chapter. As the topic of e-government progresses in Brazil, strategic monitoring and evaluation policies are also required to uncover existing gaps [11]. We believe these measures can help build a more equal and digital society.

Appendix A.

Cited references classified according to their contribution to the main area of research

Authors	Main Goal	Classification/Issue
Agune, R., & Carlos, J. [14]	The authors examine events that occurred in the context of changes in the use of information technology and the incorporation of electronic government into Brazilian public administration in the 1990s	Brazilian context

(continued)

(continued)

Authors	Main Goal	Classification/Issue
De Vaujany, F. X., Leclercq-Vandelannoitte, A., Munro, I., Nama, Y., & Holt, R. [3]	The use of technology to assist in the management of work and methods of organizing work practice	Data and algorithms for decision support system
Dias, L. N. da S., Aquino, A. C. B., Silva, P. B., & Albuquerque, F. dos S. [30]	System integration and accounting and budgeting information implementation challenges in Brazilian local governments	E-government
Engin, Z., & Treleaven, P. [8]	Examples of how GovTech systems could be used to improve public engagement and increase the efficiency of government services	Digital Government
Filgueiras, F., Moura Palotti, P. L., & Nascimento, M. I. B. 3 [24]	The authors draw on international literature on the use of evidence for the digital transformation of public services to present problems and perspectives	Digital Government
Finquelievich, S., Martínez, S. L., Jara, A., & Fressoli, M. (2004)	Through case studies of two Latin American countries, the authors show how the use of information and communication technology tools can be used to improve governance and create new forms of interaction between governments and citizens	Digital Government
Gorwa, R. [25]. What is platform governance?	The author seeks to map an interdisciplinary research agenda for platform governance in order to capture the layers of governance relationships between key parties in today's platform society, such as platform companies, users, advertisers, governments, and other political actors	Digital Government
Gil-Garcia, J. R., & Pardo, T. A. [18]	The authors present an examination of a subset of the literature on the challenges to information technology initiatives used to guide e-government efforts	E-government
Jia, T., Wang, C., Tian, Z., Wang B., & Tian, F. (2022)	The authors showed how artificial intelligence technology could aid financial decision-making	Data and algorithms for decision support system

(continued)

(continued)

Authors	Main Goal	Classification/Issue
Kellogg, K. C., Valentine, M. A., & Christin, A. [2]	The authors synthesize interdisciplinary research on algorithms at work while discussing the implementation of algorithmic technologies in organizations	Data and algorithms for decision support system
Kim, S., Andersen, K. N., & Lee, J. [23]	The authors discussed how the nature of government work has fundamentally changed and how platform government necessitates new approaches to public management	Digital Government
Laia, M. M. D., Cunha, M. A. V. C. D., Nogueira, A. R. R., & Mazzon, J. A. [12]	The authors discuss electronic government policies and actions in Brazil, highlighting the limitations of using ICT to provide integrated public services	E-government
Leite, H., Hodgkinson, I. R., & Gruber, T. [26]	The article focuses on the positive impact of telemedicine in assisting with service provision challenges and opportunities, using the Brazilian case as an example	E-government
Maynard, D. [20]	The chapter provides a historical overview of the Internet's implementation and popularization, with a focus on the situation in Brazil between 1995 and 2017	Brazilian context
Nam, H., Kim, S., & Nam, T. [1]	The paper discussed the role of the public sector and government in the future development of the ICT sector in terms of environments, regulations, policies, and governance in a number of countries around the world	Data and algorithms for decision support system
Nemer, D. [10]	The author demonstrates how residents of slum areas appropriate everyday technologies, highlighting some useful references that are relevant in addressing how Information and Communication Technologies are used in various socioeconomic contexts and approaching the digital divide and digital inclusion in Brazil	Brazilian context

(continued)

(continued)

Authors	Main Goal	Classification/Issue
Neves, F. R., & Silva, P. B. [11]	The authors provide an understanding of how government websites are used as an e-government tool in local governments, highlighting how e-government concepts have been empirically applied in some Brazilian local governments	E-government
Neves, F. R., Silva, P. B., & Carvalho, H. L. M. [9]	The study depicts the use of AI-based system artifacts to assist auditors in the Brazilian Supreme Audit Institution's surveillance against fraud and corruption	Data and algorithms for decision support system
Przeybilovicz, E., Cunha, M. A., & Meirelles, F. D. S. [21]	The study illustrates the infrastructure characteristics and use of information and communication technologies in Brazilian municipalities, with an eye toward developing e-government and smart cities initiatives	E-government
Reinhard, N., & Dias, I. D. M. [16]	The authors discuss the evolution of the Brazilian Electronic Government Program, which began in 2000, as well as an examination of the structure of the country's computerization process	Brazilian context
Saldanha, D. M. F., & Silva, M. B. D. [19]	The authors investigated the specifics of Brazil's electronic voting system, which provides society with control and oversight	Brazilian context
Santos, E. M. D., & Reinhard, N. [13]	The authors examine the use of electronic government in Brazil, using data from 2009 to 2013, to create a profile of the use of its services and their motivations for not using them	E-government
Segurado, R. [18]	The author's chapter discusses Internet governance in Brazil and its role, particularly after the approval of the regulatory framework, which includes principles, assurances, rights, and duties for Internet users and providers in the country	Brazilian context

(continued)

(continued)

Authors	Main Goal	Classification/Issue
Tambe, P., Cappelli, P., & Yakubovich, V. [5]	The authors explain why the data science community is skeptical of causally reasoning AI systems and identify four challenges to using data science techniques in human resource practices	Data and algorithms for decision support system
Teklu, F., Sumalee, A. and Watling, D. [6]	The article examines the problem of optimizing the traffic of an urban network and the performance of the algorithms using a case study of Chester, UK	Data and algorithms for decision support system
Wiens, J., Price, W. N., & Sjoding, M. W. [4]	The article examines and highlights the potential of algorithms in healthcare, as well as calls for their bias	Data and algorithms for decision support system

References

1. Nam, H., Kim, S., & Nam, T. (2022). Identifying the directions of technology-driven government innovation. *Information, 13*(5), 208. https://doi.org/10.3390/info13050208
2. Kellogg, K. C., Valentine, M. A., & Christin, A. (2020). Algorithms at work: The new contested terrain of control. *Academy of Management Annals, 14*(1), 366–410. https://doi.org/10.5465/annals.2018.0174
3. De Vaujany, F. X., Leclercq-Vandelannoitte, A., Munro, I., Nama, Y., & Holt, R. (2021). Control and surveillance in work practice: cultivating paradox in 'new' modes of organizing. *Organization Studies, 42*(5), 675–695. https://doi.org/10.1177/01708406211010988
4. Wiens, J., Price, W. N., & Sjoding, M. W. (2020). Diagnosing bias in data-driven algorithms for healthcare. *Nature Medicine, 26*(1), 25–26. https://doi.org/10.1038/s41591-019-0726-6
5. Tambe, P., Cappelli, P., & Yakubovich, V. (2019). Artificial intelligence in human resources management: Challenges and a path forward. *California Management Review, 61*(4), 15–42. https://doi.org/10.1177/0008125619867910
6. Teklu, F., Sumalee, A., & Watling, D. (2007). A genetic algorithm approach for optimizing traffic control signals considering routing. *Computer-Aided Civil and Infrastructure Engineering, 22*, 31–43. https://doi.org/10.1111/j.1467-8667.2006.00468.x
7. Jia, T., Wang, C., Tian, Z., Wang, B., & Tian, F. (2022). Design of digital and intelligent financial decision support system based on artificial intelligence. *Computational Intelligence and Neuroscience.* https://doi.org/10.1155/2022/1962937
8. Engin, Z., & Treleaven, P. (2019). Algorithmic government: Automating public services and supporting civil servants in using data science technologies. *The Computer Journal, 62*(3), 448–460. https://doi.org/10.1093/comjnl/bxy082
9. Neves, F. R., Silva, P. B., Carvalho, H. L. M. (2019). Artificial Ladies against corruption: searching for legitimacy at the Brazilian Supreme Audit Institution. *Revista de Contabilidade e Organizações, 13*, 31–50. https://doi.org/10.11606/issn.1982-6486.rco.2019.158530
10. Nemer, D. (2022). *Technology of the Oppressed: Inequity and the Digital Mundane in Favelas of Brazil.* MIT Press. https://doi.org/10.7551/mitpress/14122.001.0001

11. Neves, F. R., & Silva, P. B. (2021). E-government in local governments' websites: from visible to invisible. *Revista Catarinense da Ciência Contábil, 20*, e3160. https://doi.org/10.16930/2237-766220213160

12. Laia, M. M. D., Cunha, M. A. V. C. D., Nogueira, A. R. R., & Mazzon, J. A. (2011). Electronic government policies in Brazil: context, ICT management and outcomes. *Revista de Administração de Empresas, 51*(1), 43–57. https://doi.org/10.1590/S0034-759020110001 00005

13. Santos, E. M. D., & Reinhard, N. (2016). Serviços de Governo Eletrônico: um panorama no Brasil. In: *Artefatos digitais para mobilização da sociedade civil: Perspectivas para avanço da democracia.* Salvador, UFBA (2016). https://doi.org/10.7476/9788523218775.0010

14. Agune, R., & Carlos, J. (2005). Governo eletrônico e novos processos de trabalho. In: *Gestão pública no Brasil contemporâneo* (pp. 302–315). Fundap

15. Finquelievich, S., Martínez, S. L., Jara, A., & Fressoli, M. (2004). The social impact of introducing ICTs in local government and public services: Case studies in Buenos Aires and Montevideo. In M. Bonilla, & G. Cliche (Eds.), *Internet and Society in Latin America and the Caribbean* (pp. 147–192).

16. Reinhard, N., & Dias, I. D. M. (2005). Categorization of e-gov initiatives: a comparison of three perspectives. In: *Congreso Internacional del Clad sobre la Reforma del Estado y de la Administración Pública, Santiago, Chile.*

17. Pereira, L. C. B. (2014). Reforma da nova gestão pública: agora na agenda da América Latina, no entanto. *Revista do Serviço Público, 53*(1), 5–27. https://doi.org/10.21874/rsp.v53i1.278

18. Gil-Garcia, J. R., & Pardo, T. A. (2005). E-government success factors: Mapping practical tools to theoretical foundations. *Government information quarterly., 22*(2), 187–216. https://doi.org/10.1016/j.giq.2005.02.001

19. Saldanha, D. M. F., & Silva, M. B. D. (2020). Transparency and accountability of government algorithms: The case of the Brazilian electronic voting system. *Cadernos EBAPE BR, 18*, 697–712. https://doi.org/10.1590/1679-395120190023x

20. Maynard, D. (2018). An introduction to the history of the internet: A Brazilian perspective. In *The Internet and Health in Brazil* (pp. 15–26). Springer. https://doi.org/10.1007/978-3-319-99289-1_2

21. Przeybilovicz, E., Cunha, M. A., & Meirelles, F. D. S. (2018). The use of information and communication technology to characterize municipalities: Who they are and what they need to develop e-government and smart city initiatives. *Revista de Administração Pública., 52*, 630–649. https://doi.org/10.1590/0034-7612170582

22. Segurado, R. (2018). The Brazilian civil rights framework for the internet: A pioneering experience in internet governance. In *The Internet and Health in Brazil* (pp. 27–45). Springer. https://doi.org/10.1007/978-3-319-99289-1_3

23. Kim, S., Andersen, K. N., & Lee, J. (2022). Platform government in the era of smart technology. *Public Administration Review, 82*(2), 362–368. https://doi.org/10.1111/puar.13422

24. Filgueiras, F., Moura Palotti, P. L., & Nascimento, M. I. B. (2022). Policy Design e uso de evidências: O caso da Plataforma GOV.BR. In N. M. Koga, P. L. M. Palotti, J. Mello, & M. M. S. Pinheiro (Eds.), *Políticas públicas e usos de evidências no Brasil: conceitos, métodos, contextos e práticas*, vol. 1 (pp. 521–550). https://doi.org/10.38116/978-65-5635-032-5/capitu lo16

25. Gorwa, R. (2019). What is platform governance? *Information, Communication and Society, 22*(6), 854–871. https://doi.org/10.1080/1369118X.2019.1573914

26. Leite, H., Hodgkinson, I. R., & Gruber, T. (2020). New development: 'Healing at a distance'—telemedicine and COVID-19. *Public Money and Management, 40*(6), 483–485. https://doi.org/10.1080/09540962.2020.1748855

27. OECD. (2020). *OECD Telecommunication and Broadcasting Review of Brazil 2020.* OECD Publishing, Paris. https://doi.org/10.1787/30ab8568-en

28. OECD. (2018). Digital government review of Brazil: Towards the digital transformation of the public sector. In *OECD Digital Government Studies.* OECD Publishing, Paris. https://doi.org/10.1787/9789264307636-en

29. CGI.br—Brazilian Internet Steering Committee. (2020). Survey on the use of information and communication technologies in the Brazilian public sector: ICT Electronic Government 2019. Retrieved June 08, 2022, from https://cetic.br/media/docs/publicacoes/2/20200707094309/tic_governo_eletronico_2019_livro_eletronico.pdf
30. Dias, L. N. da S., Aquino, A. C. B., Silva, P. B., & Albuquerque, F. dos S. (2020). Outsourcing of fiscal transparency portals by municipalities. *Revista de Contabilidade e Organizações, 14*. https://doi.org/10.11606/issn.1982-6486.rco.2020.164383

Chapter 10
The Role of Public Libraries in Improving Public Literacy Through Twitter Social Media in Indonesia

Muslimin Machmud, Andi Ernie Zaenab Musa, Budi Suprapto, and Salahudin

Abstract Social media has become one of the important elements in attracting attention and communicating to the public regarding the implementation of government programs in the last decade. This opportunity was also used by the National Library of the Republic of Indonesia (Perpusnas RI) by creating social media accounts to maintain communication and provide important information to the public via Twitter. The study aims to determine the role of social media, namely Twitter @perpusnas1 belonging to the National Library of the Republic of Indonesia (Perpusnas RI), in increasing literacy. It employs qualitative descriptive research methods, namely research that describes research results more broadly with the help of NVivo 12 Plus software which is valuable and effective in assisting qualitative research efficiently. The study results display the captured data: how much intensity posts, which accounts make the most "mentions," how networks or relationships are based on usernames and hashtags, and post content focusing on literacy keywords; for example, literacy culture, literacy information, literacy familiarization, and literacy greetings.

Keywords National Library of the Republic of Indonesia · Literacy · Social Media · Twitter · Communication · Government

10.1 Introduction

Following the development of social media, governments worldwide have been working hard to identify how social media platforms can be used as effective tools for strategic communication [12]. World leaders are adopting new forms of political communication—in this case, social media (Twitter and Facebook). Not unlike the

M. Machmud (✉) · B. Suprapto
Department of Communication Science, Universitas Muhammadiyah Malang, Malang, Indonesia
e-mail: machmudmus@umm.ac.id

A. E. Z. Musa
Department of Government Studies, Universitas Muhammadiyah Malang, Malang, Indonesia

Salahudin
Politeknik Maritim AMI Makassar, Makassar, Indonesia

advent of radio or television, social media provides world leaders with a new platform to broadcast messages, mobilize constituents, and persuade citizens [3]. There is a rapid growth in the development of social media, and the complexity is decreasing day by day, resulting in more friendly apps for individuals young and old. Social media is one of the most effective tools for versatile information. These include online technologies that allow people to communicate easily over the internet and share resources [20].

Virtual communication has become universal due to the widespread use of social media, allowing citizens of cyberspace to share opinions freely. Social media use has been defined as the consumption of certain digital media that allow users to connect, communicate and interact with one another through social networking sites and instant messaging. Many social media sites are primarily designed to facilitate communication between individuals and groups, for example, Twitter, LinkedIn, Microblogs, WhatsApp, Line and WeChat, and others [9]. The newer generation of social media applications that will be based on Web3 platform will even provide features like data privacy and monetization of data that will be shared on these platforms. Virtual worlds such as Metaverse will use these Web3 platforms to socialize the data among the different variety of users.

The ubiquity of social media has a profound effect on how we communicate and is vital to society and business. Social media tools have helped break down geographic barriers that once limited communication and have led to an explosion of electronic participation, virtual presence, and online communities. Professional benefits of social media include information sharing, publicity, and giving and receiving support and advice. People browse and contribute to their social media accounts regularly using smart devices; some prefer to communicate using social media rather than participating in face-to-face interactions. Social media tools also allow citizens to share advice and information with their local communities, from promoting events to searching for lost pets and assisting government engagement with citizens [10].

Today's social media users have expanded, from individuals, companies, and brands that use social media to public policymakers (for example, governments regulators). Past research has also shown how the Chinese government strategically uses millions of online comments to distract the Chinese public from discussing sensitive issues and promoting nationalism. Social media as a political tool may be more influential in a region where the government is notoriously controlled and censored mainstream media. Social media users have captured important changes in the social media space through the lens of important stakeholders, including consumers, industry/practices, and public policy [2].

The ability to search, select, sort, and use information in the digital world is known as media literacy. Good media literacy skills have the potential to produce quality information. This social media literacy shows the ability to filter information found through social media and use it for beneficial things to the lives of users. Healthy use of social media begins with social media literacy from social media users [26]. Furthermore, media literacy—a peer influence approach can help address the unique context of social media in which peer interactions and highly visual, accessible formats create performance pressure for young people. Finally, media literacy aims to increase critical thinking and skepticism about the media and increase proficiency in media building to reduce its persuasive influence [21].

Literacy is a special skill level needed in everyday life and is a general term for a set of skills and individual abilities in reading, writing, speaking, calculating, and solving problems. Therefore, literacy and language skills cannot be separated. According to UNESCO, human understanding is influenced by academic abilities, national contexts, institutions, cultural values, and experiences. According to the Big Indonesian Dictionary, literacy is reading and writing. However, the meaning of literacy is understood as the ability to read and write and has a more complex and dynamic understanding. The notion of 'critical digital literacy' requires rethinking, given the rapidly changing nature of young people's digital practices. In the contemporary era, the success of young people as students who will be engaged as citizens and employees of the future has been associated with 'digital literacy'. Some theorists claim that without the skills to use and evaluate digital tools now found in most informal and formal contexts, young people will be left behind in various aspects of their lives—from work to social interactions [23]. It is important to create social media campaigns on the some the upcoming technologies to catch more eyes and attention span. As companies like Facebook, Google and Microsoft moves towards building metaverses, there is an opportunity for the social media platforms to swith on these virtual worlds as well and create a unique branding with the audience that they wish to connect.

The literature covering the field of media literacy has been described as "complex ideas", or "a citizen's ability to access, analyze, and generate information for a particular outcome". Educating people in these discussions aims to create informed and independent citizens who question the information they receive, value aesthetics, develop self-esteem and competence and have a sense of advocacy. Concern over fake news has sparked renewed interest in various forms of media literacy. The general expectation presupposes that literacy interventions help the audience be educated to avoid the harmful effects of misleading information [15]. The actions of policymakers and civil society organizations to increase media literacy can also influence the observed trends [1].

Follow-up as a part of media literacy requires the National Library of Indonesia to participate in society's current trend, which is in a period of dependence on social media, whatever it may be. The need for the involvement of the National Library of Indonesia as one of the central institutions that have the burden of educating the lives of the Indonesian people through breakthroughs or millennial programs is an effort to adapt to the socio-cultural conditions of today's society. Thus, the steps taken by the National Library of Indonesia in keeping pace with the times are by utilizing social media as an effort to promote literacy to the Indonesian people today, one of which is Twitter. Thus, the focus of the research is to see and analyze the role of the National Library of Indonesia"s Twitter social media as a means for literacy communication to the public.

10.2   Literature Review

10.2.1   Social Media as a Means of Communication and Information

Social media consists of communication websites that facilitate relationships between users from different backgrounds, resulting in a rich social structure. When we refer to social media, apps like Facebook, WhatsApp, Twitter, YouTube, LinkedIn, Pinterest, and Instagram often come to mind. The app is driven by user-generated content and is highly influential in various settings, from buying/selling behavior, entrepreneurship, political issues to capitalism. In recent years user-generated content from social media has become one of the most important information channels across public administration and political contexts. Previous studies have also shown that the use of social media increases public engagement and transparency. Social media technology is no longer considered a platform for socialization and meetings but is recognized for its ability to encourage aggregation. Facebook, online communities, and Twitter are the three most popular networks targeted by publications in social media research [16].

During public crises, governments must act quickly to communicate crisis information effectively and efficiently to public members; Failure to do so will surely make residents afraid, uncertain, and anxious about the existing conditions. Through engagement with the public, governments can develop citizens' understanding of their actions and resilience in responding to crises and improve government agencies' ability to process crisis information and provide public services. Due to its open, dialogical, and participatory nature, social media offers significant benefits in delivering synchronous and interactive communication between government and citizens, bringing new impetus to citizen engagement. Government agencies worldwide are enthusiastically exploring the use of social media to encourage citizen engagement in crisis management [8].

In recent years, social media platforms have become a highly leveraged source for communication with many benefits realized, for example, fast response, information reach, and low costs and risks, for example, spam, false information, and rumors as a consequence. At the same time, effective analysis of the data generated by social media can provide useful insights in communication to manage situations more effectively [28].

10.2.2   Social Media as a Means of Communication and Public Policy Information

Social media and other types of online media are becoming popular platforms where interactions as diverse as information sharing and civic engagement in public administration occur. Such interactions through online platforms contribute to overcoming the information asymmetry between the government and citizens, thereby increasing the government's responsiveness. If political leaders actively use social media in public administration and policy emerge, their intermediary role on social media will likely help overcome difficulties in increasing responsiveness. The research findings show that the mayor of Seoul plays the most important role as a liaison on the Twitter network. In particular, the mayor serves as a bridge between different groups of citizens and public officials as well as a hub for the most connected users in the network, contributing to increasing government responsiveness by enabling it to overcome disconnects between citizens and local government, and information asymmetry between mayors, public officials, and citizens [11].

Social media plays an essential role in modern society. Over the past decade, governments have also adopted this tool as a new method of communication and engagement with their citizens. The research results by [5] show that most Andalusian local governments have official corporate Twitter accounts with a certain level of activity. Content related to the city's cultural activities and promotions was frequently tweeted. Retweets are the most frequent way for citizens to interact with city council accounts regarding citizen engagement. Involvement does not appear to be related to the municipality's population. Research findings also show that photos and videos generate more engagement (retweets and favorites) than other media types. The same was true for content types where sports-related tweets (retweets and favorites) and environmental issues (replies) resulted in the highest engagement.

Despite its lack of resources, the findings shed light on how Vietnam has demonstrated political readiness to combat the emerging pandemic from the start. Timely communication about the development of an outbreak from the government and the media, combined with current research on the new virus by the Vietnamese science community, has provided a reliable source of information. It emphasizes the need for immediate and genuine cooperation between governments, civil society, and private individuals. This case study offers valuable lessons for other countries regarding the simultaneous struggle against the COVID-19 pandemic and the overall response to the public health crisis [19].

10.2.3   Social Media as a Means of Library Literacy

The National Association for Media Literacy Education (NAMLE) holds the third U.S. Media Literacy Week. In addition, the American Library Association announced their new partnership with Stony Brook University to create "Media Literacy in

Your Library". Media literacy is most often described as a set of skills that promote critical engagement with the messages generated by the media. Fundamentally, media literacy is "the active investigation and critical thinking of the messages we receive and create," and most proponents emphasize this relationship to critical thinking. The U.S. The National Association for Media Literacy Education (NAMLE) defines media literacy as "the ability to access, analyze, evaluate, create, and act using all forms of communication" [7].

Social media is a means for libraries to provide literacy education to the public. Literacy-related activities and information can be broadcast on social media. Immediately, the information displayed can be accessed by thousands of people either through computers, laptops, or cell phones, which currently have more users than the total population of Indonesia. Various social media applications can be used to learn through various information, data and issues contained in them. On another aspect, social media is also a means to convey various information to other parties. Social media contents come from various parts of the world with various backgrounds (33) cultural, social, economic, beliefs, traditions, and tendencies [22].

The development of computer technology, telecommunications, information has a massive impact on the library. This development gave rise to various new names in the concept of modern libraries, such as libraries without walls (Library without walls), digital libraries (digital libraries), virtual libraries (virtual libraries), and electronic libraries (e-library) [14]. The development of the internet and the rapid development of new sources of information require libraries to make changes, both in collections and in terms of service patterns. Libraries considered a source of knowledge have an important role in providing knowledge for library users in utilizing digital information both in learning and having an impact on life in general. A Digital library is the application of information technology used to obtain, store, and disseminate information packaged in digital form. In simple terms, it can be said as a place to store library collections in digital form.

Digital libraries make it easy for users to access electronic information sources with tools that are easily accessible in limited time and opportunities. Users are no longer physically bound to library service hours, where users have to visit the library directly to get information. Therefore, digital libraries can be used as a means and solution to the problem of limited access. The development of a digital library is one strategy that can be optimized in response to people's low interest in reading because they tend to stick to their gadgets. Although not the primary key, the existence of a digital library is quite influential in developing interest in reading and information to the public. Libraries as digital media act as transmission media or communication channels that convey messages (digital information) to community users through digital technology. The concept of digital libraries is not limited to converting the books to digital images. There are apps and forums where a person can get audio book with focus on key aspects of the book for easy learning. This concept is getting extended with video books having the visual representation of the ideas expressed in the book. Human mind is able to capture audio and video much faster then a written text and hence these new innovations help to more audience and better learning experience.

Libraries as a source of knowledge have an important role in educating library users to utilize digital information both in learning and impacting life in society. The development of information literacy materials is multi-literacies, which includes literacy such as digital technology, information, multi-media, visual (images), audio, critical thinking, and understanding of ethical, moral, legal, social, cultural issues that surround the digital environment and how to participate in online communities politely and responsibly. The generation of digital natives still needs traditional literacy cultures such as writing, reading, and listening to improve digital literacy skills. Librarians need to have creativity in designing interesting information literacy activity programs for their users by following the trend of library users currently dominated by the digital generation group [25].

10.3   Research Method

The study employs NVivo 12 Plus, a qualitative data analysis software. This study uses Qualitative Data Analysis (QDA) with the computer programming assistant Nvivo 12 Plus [24]. Analyzes were carried out using NVivo 12 Plus, a qualitative data analysis software that facilitates the collection, categorization, mapping, analysis, and visualization of qualitative data, including those collected from documents (memos, reports, legislation, and photographic documents) and through interviews [4] in [27]. In this study, the documents collected are the activities of one of the Twitter accounts that become the object of this study. This research was carried out by collecting the activity of the Twitter account @perpusnas1 using the capture feature (Google N-capture). The data taken is related to the activity of the @perpusnas account in informing literacy programs at the National Library of Indonesia. Data collection starts from July 2019 to January 2022.

10.4   Findings

The emergence of social media has changed the communication landscape of organizations, enabling organizations to build symmetrical two-way communication programs. Therefore, organizations will use social media to communicate their efforts effectively and, thereby, build public trust, and that social media can offer new opportunities for transparency and interactivity with stakeholders, which in turn can trigger positive evaluations of the image and organizational reputation, which generates public trust in both the short and long term [17]. Furthermore, [18] also stated that social media is considered to have high effectiveness in conveying information to the public.

The National Library of Indonesia's Twitter social media account has the username @perpusnas1, which will be the main topic in this discussion. It is necessary to note that the @perpusnas1 account has joined Twitter since 2012. So it has been almost ten years that the @perpusnas1 account has existed on Twitter as one of the official accounts of the Indonesian National Library of Indonesia to make it easier for the public to communicate and make it easier for the Indonesian National Library to become an informative institution related to with efforts to improve literacy skills for all Indonesian people.

Social media accounts belonging to Perpunas RI have an important role in always being regular and intense in conveying various information, education, and easy access for the public. The more often the account is active on social media, especially to provide education to the public, the literacy communication strategy from the National Library of Indonesia to the community is considered successful. This success is also supported by the high involvement of the public in responding to any information or Twitter activity of the @perpusnas1 account through feedback or responses to certain messages or content being shared. Thus, it is important to know how often the @perpusnas1 account is active and can grab the attention of its followers or other users (Fig. 10.1).

Based on the graph above, the density or intensity of postings or social media activities by the Twitter account @perpusnas1 did not experience either a periodic increase or a periodic decrease. The data shows that the activity of the @perpusnas1 account has increased and decreased activity randomly and not neatly systemized. It can be seen that the most social media activity carried out by the @perpusnas1 account occurred in March 2021. On the contrary, the slightest activity occurred in July 2019. Then if you look at the comparison of the number of posts or tweet activity per day, the number of tweet activities in March 2021 reached an average of seven times activity, and the average daily tweet activity in July 2019 did not reach one per day.

The results obtained based on the image above illustrate how many activities were carried out by the Twitter account belonging to the National Library of Indonesia within 30 months, starting from July 2019 to January 2022. During those 30 months, it appears that the Twitter account @perpusnas1 experienced fluctuation or inconstancy. Fluctuations illustrate that the activity of Twitter accounts carried out has increased and decreased that was not fixed. This activity shows that the communication process of the National Library of Indonesia through social media aimed at its followers or other users has experienced inconsistencies. It means that no policy requires intense communication with users and followers based on the number of posts per month.

The next discussion is about activities that also show the relationship or network of the @perpusnas1 Twitter account activities with other users and their followers, namely by using the "mention" feature. This feature is an activity where the @perpusnas1 account mentions usernames from other accounts in its tweet activity. This activity is important to analyze to find out the users who are actively involved in social media activities with the @perpusnas1 account and show how active the @perpusnas1 account is in communicating with their followers or other users (Fig. 10.2).

The data above is formed as a result of processing based on the value of the majority who carry out "mention" activities or "names" with the Twitter account @perpusnas1 from a predetermined period of time, which is for 30 months as previously discussed.

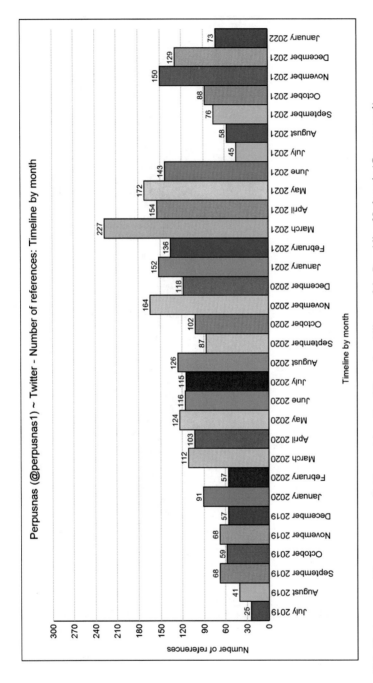

Fig. 10.1 The intensity of literacy posts on Twitter social media of the national library of the Republic of Indonesia (@perpusnas1)

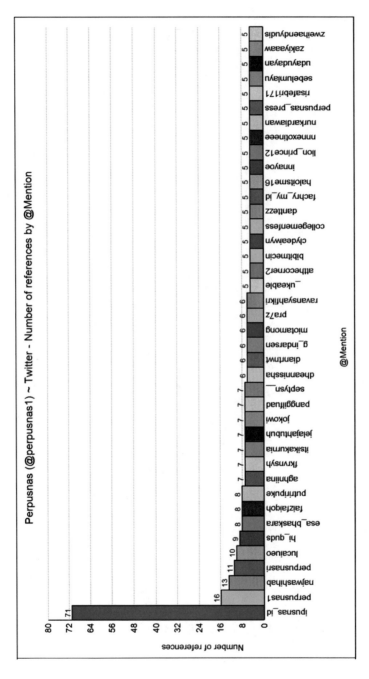

Fig. 10.2 Mention literacy on social media Twitter national library of the Republic of Indonesia (@perpusnas1)

Based on the picture above, it can be seen that the activity "mention" which is mostly done by @perpusnas1 is with the @ipusnas_id account. Furthermore, most of the recorded activities share synopsis information from a book to inform each other regarding service systematics at the National Library of Indonesia through the IPUSNAS application facilitated by the Twitter medium. The image above also shows that the "mention" activity, whether carried out by the @perpusnas1 account or vice versa, targets all follower or user accounts—not limited to large accounts belonging to agencies or individual officials—who gave good responses related to bro. These infographics are distributed to the systematic flow of services provided.

The following discussion will be the literacy network of the @perpusnas1 Twitter account and several other accounts for the similarity of words or something being discussed. In this case, all user activities, both "mentions," "likes," to "re-tweets," are the aspects behind getting the results as shown below. The @perpusnas1 account carries out the more frequent interaction activities, and it will show how the @perpusnas1 account tries to maintain effective communication with its followers or other Twitter users who are interacting with the @perpusnas1 account through mentions, "likes," to "re-tweets" (Fig. 10.3).

The communication network based on the data results above shows several links or ties between one account and another. The bond between the user account names indicates that there has been a communication relationship between actors. The relationship from the picture above is based on the similarity of words in every tweet activity carried out between linked accounts. Based on the figure above, the accounts with the thickest ties or lines with the @perpusnas1 account are the @ipusnas_id, @20detik, and @detikcom accounts. If seen, the three accounts are official accounts of large organizations in Indonesia, namely the @ipusnas_id account, one of the programs belonging to the National Library of Indonesia. This software-based digital library application can be accessed on various types of smartphones. Then two other accounts that also interact a lot with the @perpusnas1 account are @20detik and @detikcom, which are official social media accounts belonging to one of Indonesia's largest media and news companies.

The use of hashtags is an activity that should not be missed. For this reason, the following discussion is related to the relation of using hashtags in every tweet activity carried out by the @perpusnas1 account. Hashtags can be used on Twitter to attach tweets to broader discussions and allow other Twitter users to follow specific topics and related hashtags. Twitter allows to respond to or invite certain other users by adding an @ marker before the username of the targeted user, retweet messages created by other Twitter users, and tag messages using hashtags (with a # sign) and share links to websites. Hashtags have been used to select data sets for analysis and to identify ad-hoc publics on Twitter [13] (Fig. 10.4).

The picture above also shows how the relation and connection between hashtags in tweeting activities carried out by the National Library. The picture above shows how one hashtag can be connected to another hashtag as long as they have similarities or similarities in supporting words or words contained in hashtag writing. Some of the most frequently mentioned hashtags are #Friends of the National Library, a #library, #librarynational, #serviceperpusnas, and #National Library. The mention of hashtags

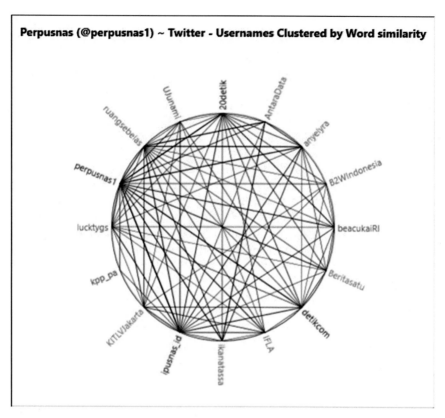

Fig. 10.3 National library of Indonesia literacy network on Twitter social media (@perpusnas1). Based on usernames

in every tweet made by the @perpusnas account is intended to help make it easier for followers or the general public of Twitter users to find out some information related to the National Library of Indonesia.

Based on the previous discussion, it is important to note that as an official social media belonging to a state institution, it is important to include all information on program activities in its official social media. Information on the program of activities is intended as an institutional effort to maintain the institution's quality and level of transparency to the public. So, the following discussion is about how the @perpusnas1 account preserves content and information about important aspects related to efforts to improve the literacy quality of the Indonesian people through several keywords, such as literacy culture, information literacy, grounding literacy, and literacy greetings.

Literacy culture is a human habit or tendency to realize their writing, reading, and thinking abilities. Literacy culture is important to be developed to improve the quality of life of human resources in a country, as well as in Indonesia. Literacy culture is one of the main topics of discussion in writing this article to know how

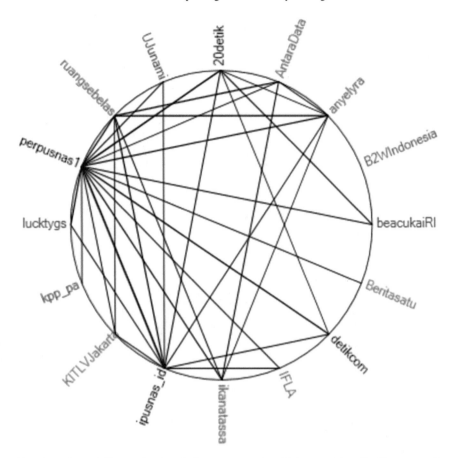

Fig. 10.4 National library of Indonesia literacy network on Twitter social media (@perpusnas1). Based on hashtags

much coverage of literacy culture content is published by the @perpusnas1 account (Fig. 10.5).

The results above show how the activity of the @perpusnas1 account in providing literacy content to followers and other users. The data above illustrates that the content provided by the @perpusnas1 account is trying to invite its followers and other users in webinar activities organized by the National Library of Indonesia, accompanied by several steps of the registration flow. Furthermore, the theme carried in the webinar focuses on efforts to transform and develop libraries to build superior and advanced human resources so that the hope in the future is to be able to realize community welfare and economic recovery and increase knowledge for the community (Fig. 10.6).

Based on the data generated from the image above, the content on literacy based on the topic "Information Literacy" shows how the @perpusnas1 account seeks to inform and invite users and librarians to continuously increase their scientific capacity and knowledge by utilizing the digital features that the National Library of

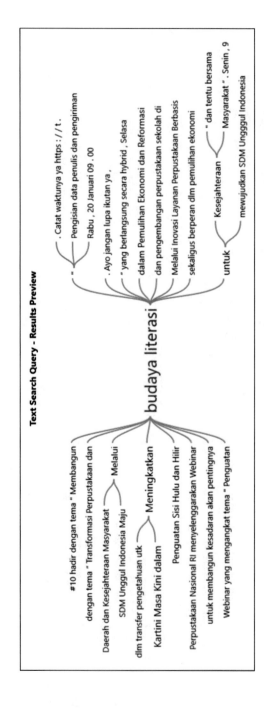

Fig. 10.5 Literacy content "Budaya Literasi/Culture of Literacy" National Library of the Republic of Indonesia on Twitter Social Media (@perpusnas1)

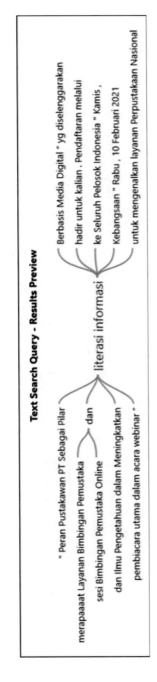

Fig. 10.6 Literacy content "Literasi Informasi/Literacy Information" national library of the Republic of Indonesia on Twitter social media (@perpusnas1)

Indonesia has. RI and digital media belong to each librarian. Therefore, the effort made by the @perpusnas1 account in campaigning or promoting the use of digital media belonging to the National Library of Indonesia is an effort to instill information literacy in the readers of the content and the targets of the forum (Fig. 10.7).

Based on the above data processing results, the content related to "earth literacy" contains information about organizing a writing competition or writer's festival with the theme "Indonesian Reading Ambassador Literacy Safari." Furthermore, it should be noted that to promote literacy, which can be understood from the data above, there are activities to support and improve literacy skills for the participants of this writing competition, so it can be said that efforts to promote literacy are one of the proper steps to be informed and disseminated through accounts @perpusnas1 (Fig. 10.8).

Related to literacy content that has a lot of involvement and mention is "Salam Literacy," one of the slogans of the National Library of Indonesia, which is often echoed and mentioned through the tweet of the @perpusnas1 account. Most of the tweet activities that include the mention of "Salam Literasi" are intended to greet followers of the @perpusnas1 account, interact with followers or other users, provide support for activity programs that are still in the family of interests to campaign for the spirit of literacy, as well as words of encouragement. Moreover, motivation to always get used to or cultivate reading and thinking.

Finally, the following important discussion to discuss and analyze is how and what the @perpusnas1 account publishes forms of literacy content. In Ref. [6] states that social media has existed to present various types of information to the public, especially to its users, both related to political issues or the current situation, which is considered to be able to increase their new knowledge insight so that they have an interest in being involved in discussing social and political issues. Based on this statement, the dissemination of information to all sections of citizens is important to consider (Fig. 10.9).

The picture above is a form of activity for the @perpusnas1 account to disseminate and inform the implementation of programs that support literacy improvement campaigns. The action or activity of posting information on the program's implementation is also included in information literacy. The @perpusnas1 account stimulates followers or other Twitter users who get information about these activities to follow up on informed activities or disseminate the information. The steps of the @perpusnas1 account in informing several activities as shown in the example above can be considered an effort to communicate the spirit of literacy to the community.

10.5 Discussion

The intensity of postings and social media activities carried out by the official social media accounts of the National Library of Indonesia illustrates how the daily activities of the National Library of Indonesia's Twitter accounts communicate with their followers. In this case, the role of social media Twitter is as an effort to inform and disseminate information to educate the public. So to maintain and preserve literacy

Fig. 10.7 Literacy content "Membumikan Literasi/Establishing Literacy" national library of the Republic of Indonesia on Twitter social media (@perpusnas1)

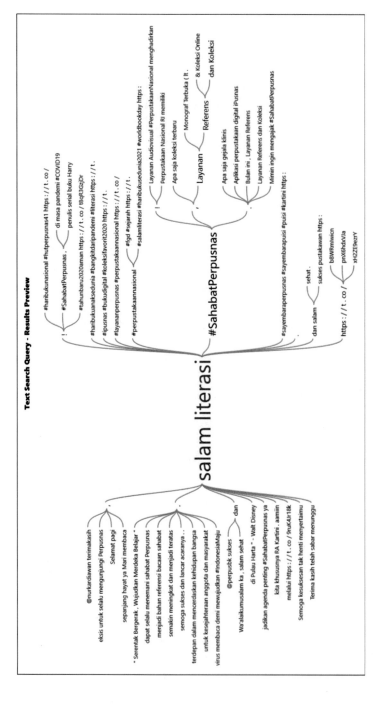

Fig. 10.8 Literacy content "Salam Literasi/Greetings of Literacy" national library of the Republic of Indonesia on Twitter social media (@perpusnas1)

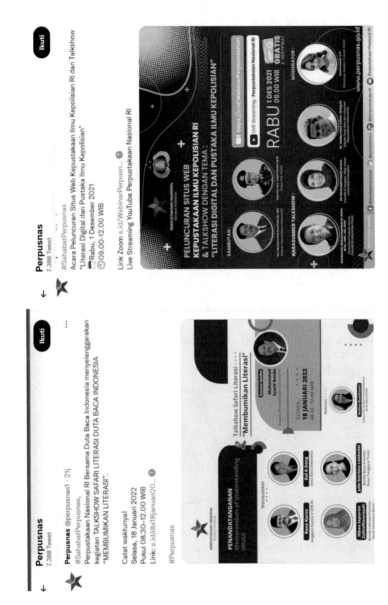

Fig. 10.9 Examples of literacy content of the National Library of the Republic of Indonesia on Twitter Social Media (@perpusnas1)

among people who are increasingly understanding the use of social media, the role of Twitter as a means of literacy for the community must be encouraged.

First, there is a fluctuating activity pattern related to the intensity of posts on the Twitter account @perpusnas1. This pattern shows how the management of the @perpusnas1 Twitter account does not have a target or plan, so the value and number of posts on its Twitter account increases periodically. The search results also show that although the @perpusnas1 account has many followers, the number of follower engagement or participation seen from the number of "reply, retweet, and like" also does not show high follower participation.

Second, the discussion regarding literacy mentions shows that the @perpusnas1 account gets the spotlight because it is one of the topics in a posted tweet. One official account carries out most of the "mention" activities. It is also supervised by the National Library of Indonesia, namely the @ipusnas_id account, which is an account that bridges information to IPUSNAS application users if they experience some problems or respond to exciting information that is being discussed, just like when they are good @perpusnas1 or @ipusnas_id accounts post related reading books on the IPUSNAS platform.

Third, relations or network usernames have the same activity as the @perpunas1 account. The visible results show that the @perpusnas1 account is still involved and is in the zone belonging to its environment. That is because the interaction between usernames shows how broad the scope of the interaction of an account with other accounts is. Because the scope of interaction activities is only limited to official accounts belonging to government organizations, the interaction pattern formed is considered less dynamic. However, it can also be judged that this step is intended so that all post information published between the two parties minimizes the expression or view of fake news.

Fourth, hashtag relationships or networks are generated from each hashtag linked or linked together because they have the same editorial sentences in each post or the same topic in each post. As written in the hashtags #Friends of the National Library of Indonesia, a #library, #librarynational, #serviceperpusnas, and #PerpusnasNasional, each of which has the same topic of sentence editorial about library service questions, activities, or events being held by the library, the obstacles experienced by the library. IPUSNAS users and other interesting information posted by the @perpusnas1 account.

Finally, posts containing cultural literacy content, information literacy, grounding literacy, and literacy greetings need to be analyzed how the core of delivery and how the @perpusnas1 account is delivered in attracting attention to the participation of followers and other users in following up on the information provided. Especially about the focus on preserving literacy in the community. The captured images in the analyzed documents show that the efforts of the National Library of Indonesia in developing, maintaining, and preserving literacy have begun to be adapted to the needs of today's society. The development of digital methods in implementing each program is a program to maintain community enthusiasm for continuing to hone their literacy skills through efficiency and effectiveness, supporting software to read, write, and think anywhere and anytime.

10.6 Conclusion

The use and utilization of social media have become an inseparable part of the government's efforts to form policies adapted to the socio-cultural conditions experienced by the people. These efforts are none other than a strategy to reach the needs of all levels of society. The high use of social media Twitter among the public must make the government also get involved in using Twitter to attract people's attention. This step has made the @perpusnas1 account a means to maintain the literacy preservation process among the people who have moved towards an increasingly open era.

The use of Twitter to maintain and preserve literacy culture among the public is also the main point in writing this article. Although it has been running for almost ten years in managing activities on Twitter, it seems that seriousness in maintaining the quality of posts and maintaining engagement or public interest still needs to be improved. Finally, equitable interaction with other users is also sought to reach users who have difficulty accessing services at the National Library of Indonesia, as part of implementing information literacy by the National Library of Indonesia.

References

1. Allcott, H., Gentzkow, M., & Yu, C. (2019). Trends in the diffusion of misinformation on social media. *Research and Politics, 6*(2). https://doi.org/10.1177/2053168019848554
2. Appel, G., et al. (2020). The future of social media in marketing. *Journal of the Academy of Marketing Science, 48*(1), 79–95. https://doi.org/10.1007/s11747-019-00695-1
3. Barberá, P., & Zeitzoff, T. (2018). The new public address system: Why do world leaders adopt social media? *International Studies Quarterly, 62*(1), 121–130. https://doi.org/10.1093/isq/sqx047
4. Bazeley, P., & Richards, L. (2013). The NVivo qualitative project book. *Journal of Chemical Information and Modeling, 53.*
5. Bonsón, E., Perea, D., & Bednárová, M. (2019). Twitter as a tool for citizen engagement: An empirical study of the Andalusian municipalities. *Government Information Quarterly, 36*(3), 480–489. https://doi.org/10.1016/j.giq.2019.03.001
6. Boulianne, S. (2019). Revolution in the making? Social media effects across the globe. *Information Communication and Society, 22*(1), 39–54. https://doi.org/10.1080/1369118X.2017.1353641
7. Bulger, M., & Davison, P. (2018). The promises, challenges, and futures of media literacy. *Journal of Media Literacy Education, 10*(1), 1–21. https://doi.org/10.23860/jmle-2018-10-1-1
8. Chen, Q., et al. (2020). Unpacking the black box: How to promote citizen engagement through government social media during the COVID-19 crisis. *Computers in Human Behavior, 110*(April), 106380. https://doi.org/10.1016/j.chb.2020.106380
9. Cheng, X., Fu, S., & de Vreede, G. J. (2017). Understanding trust influencing factors in social media communication: A qualitative study. *International Journal of Information Management, 37*(2), 25–35. https://doi.org/10.1016/j.ijinfomgt.2016.11.009
10. Dwivedi, Y. K., et al. (2018). Social media—the good, the bad, the Ugly. In *IDIMT 2013—Information Technology Human Values, Innovation and Economy, 21st Interdisciplinary Information Management Talks, 20*(March), pp. 419–423.
11. Eom, S. J., Hwang, H., & Kim, J. H. (2018). Can social media increase government responsiveness? A case study of Seoul, Korea. *Government Information Quarterly, 35*(1), 109–122. https://doi.org/10.1016/j.giq.2017.10.002

12. Eriksson, M. (2018). Lessons for crisis communication on social media: A systematic review of what research tells the practice. *International Journal of Strategic Communication, 12*(5), 526–551. https://doi.org/10.1080/1553118X.2018.1510405
13. Hellsten, I., & Leydesdorff, L. (2020). Automated analysis of actor–topic networks on twitter: New approaches to the analysis of socio-semantic networks. *Journal of the Association for Information Science and Technology, 71*(1), 3–15. https://doi.org/10.1002/asi.24207
14. Jiyanto, J., & Miftah, M. (2018). PEMANFATAN MEDIA SOSIAL PADA PERPUSTAKAN KOTA LITERASI (Studi Kasus Perpustakaan Daerah Kabupaten Sragen). *LIBRARIA: Jurnal Perpustakaan, 5*(1), 199. https://doi.org/10.21043/libraria.v5i1.2367
15. Jones-Jang, S. M., Mortensen, T., & Liu, J. (2021). Does media literacy help identification of fake news? Information literacy helps, but other literacies don't. *American Behavioral Scientist, 65*(2), 371–388. https://doi.org/10.1177/0002764219869406
16. Kapoor, K. K., et al. (2018). Advances in social media research: Past, present and future. *Information Systems Frontiers A Journal of Research and Innovation, 20*, 531–558.
17. Kollat, J., & Farache, F. (2017). Achieving consumer trust on Twitter via CSR communication. *Journal of Consumer Marketing, 34*(6), 505–514. https://doi.org/10.1108/JCM-03-2017-2127
18. Kurniawan, D., et al. (2021). Analysis of the anti-corruption movement through twitter social media: A case study of Indonesia. In *International Conference on Advances in Digital Science* (pp. 298–308). Springer. https://doi.org/10.1007/978-3-030-71782-7_27
19. La, V. P., et al. (2020). Policy response, social media and science journalism for the sustainability of the public health system amid the COVID-19 outbreak: The vietnam lessons. *Sustainability (Switzerland), 12*(7). https://doi.org/10.3390/su12072931
20. Latif, M. Z., et al. (2019). Use of smart phones and social media in medical education: Trends, advantages, challenges and barriers. *Acta Informatica Medica, 27*(2), 133–138. https://doi.org/10.5455/aim.2019.27.133-138
21. McLean, S. A., et al. (2017). A pilot evaluation of a social media literacy intervention to reduce risk factors for eating disorders. *International Journal of Eating Disorders, 50*(7), 847–851. https://doi.org/10.1002/eat.22708
22. Mustofa, M. (2017). PROMOSI PERPUSTAKAAN MELALUI MEDIA SOSIAL: Best Practice. *Publication Library and Information Science, 1*(2), 21. https://doi.org/10.24269/pls.v1i 2.691
23. Pangrazio, L. (2016). Reconceptualising critical digital literacy. *Discourse, 37*(2), 163–174. https://doi.org/10.1080/01596306.2014.942836
24. Purnomo, E. P., et al. (2021). How public transportation use social media platform during Covid-19: Study on Jakarta public transportations' twitter accounts? *Webology, 18*(1), 1–19. https://doi.org/10.14704/WEB/V18I1/WEB18001
25. Mardina, R. (2017). Literasi digital bagi generasi digital natives. Prosiding Conference Paper (Mei), pp. 340–352.
26. Safira, F., & Irawati, I. (2020). Hubungan Literasi Media Sosial Pustakawan Perguruan Tinggi dengan Kualitas Pemanfaatan e-Resources Perpustakaan. *Lentera Pustaka: Jurnal Kajian Ilmu Perpustakaan, Informasi dan Kearsipan, 6*(1), 1–12. https://doi.org/10.14710/lenpust.v6i1. 25325
27. Salahudin, S., Nurmandi, A., & Loilatu, M. J. (2020). How to design qualitative research with NVivo 12 plus for local government corruption issues in Indonesia? *Jurnal Studi Pemerintahan, 11*(3). https://doi.org/10.18196/jgp.113124
28. Stieglitz, S., et al. (2018). Sense-making in social media during extreme events. *Journal of Contingencies and Crisis Management, 26*(1), 4–15. https://doi.org/10.1111/1468-5973.12193

Printed in the United States
by Baker & Taylor Publisher Services